The People of the River

OSCAR DE LA TORRE

The People of the River

Nature and Identity in Black Amazonia, 1835–1945

The University of North Carolina Press *Chapel Hill*

Set in Arno Pro by Westchester Publishing Services

The University of North Carolina Press has been a member of the Green Press Initiative since 2003.

Library of Congress Cataloging-in-Publication Data
Names: De la Torre, Oscar (Historian), author.
Title: The people of the river : nature and identity in black Amazonia, 1835–1945 / Oscar de la Torre.
Description: Chapel Hill : University of North Carolina Press, [2018] | Includes bibliographical references and index.
Identifiers: LCCN 2018008637| ISBN 9781469643236 (cloth : alk. paper) | ISBN 9781469643243 (pbk : alk. paper) | ISBN 9781469643250 (ebook)
Subjects: LCSH: Blacks—Brazil—History. | Blacks—Race identity—Brazil. | Blacks—Brazil—Economic conditions. | Peasants—Brazil—History. | Slavery—Brazil—History. | Amazon River Region—History.
Classification: LCC F2659.N4 D387 2018 | DDC 305.896/081—dc23
LC record available at https://lccn.loc.gov/2018008637

Cover illustrations: Photo of Guilhermo from Otille Coudreau, *Voyage au Cuminá: 20 Avril 1900–7 Septembre 1900* (Paris: Lahure, 1901), 9; image of slaves collecting turtle eggs from Johann Baptist von Spix and Karl Friedrich Philipp von Martius, *Atlas zur Reise in Brasilien* (Munich: M. Lindauer, 1823), 27.

Portions of chapter 6 were previously published in a different form as "'The Land Is Ours and We Are Free to Do All that We Want': Quilombos and Black Rural Protest in Amazonia, Brazil, 1917–1929," *Latin Americanist* 56 (2012): 33–56. Used here with permission.

Per la Irene, el Pau, i el Jun. Vet aquí el perquè de tantes hores!

Para la mama, la yaya, las tatas, y los anti-Cueva.

Contents

Figures, Graphs, Maps, and Table

Acknowledgments

Writing a book takes years, and life happens while you do it. My father and my grandmother passed away during the early stages of this project, but both were key to it through their continuing support, their financial help, and their concerns about my future. I hope that my love and my gratitude will reach them wherever they are. Fortunately my wife Irene and my sons Pau and Jun joined me and remained by my side as I wrote it, so it is to them and to my mom and sisters that I dedicate this book. My sons think that the book I was writing for years was some kind of imaginary friend. By now I hope they will believe me.

This is a genuinely Atlantic book: parts of it were written in the United States, Spain, and Brazil. Javier Laviña, Miquel Izard, Gabriela Dalla Corte, Pilar García-Jordán, and Meritxell Tous are guilty of turning me into a Latin Americanist. José Luís Ruíz-Peinado provided contacts, guidance, and advice when I visited Pará for the first time in 2004; I returned there many times since. Ramón Flecha suggested that I attend the University of Pittsburgh, where Reid Andrews provided consistent and dedicated guidance throughout the years. Alejandro de la Fuente was also a demanding reader and an excellent dinner guest, although for some strange reason he did not always appreciate my jokes. I also benefited from the intellectual insight and the stimulating environment provided by Lara Putnam, Rob Ruck, Van Beck Hall, Bill Chase, Irina Livezeanu, John Markoff, Jerome Branche, John Frechione, and many others. Kavin Paulraj, Lars Peterson, Jake Pollock, Tasha Kimball, Alejandra Boza, Bayete Henderson, Julien Comte, Kenyon Zimmer, Alyssa Ribeiro, and Niklas Frykman all made the trip a funnier one. Francis Allard, Adriana Maguiña, Adolfo García, Júlia Romero, Marifélix Cubas, and Russ Maiers also walked with me along the road.

The scholars of Belém do Pará, Brazil, were exceedingly warm and welcoming. I am especially grateful to Nilma Bentes and the turma do CEDENPA, Rafael Chambouleyron, José Maia Bezerra Neto, Pere Petit, Fernando Arthur de Freitas Neves, Sara Alonso, Didier Lahon, Dionisio Poey Baró, Karl Arenz, and Daniel Barroso. We often debated about history and about politics, and had drinks together as members of O Clube. They welcomed me into their homes and, most importantly, into their lives.

A number of guides and hosts lent me a hand as I researched this book in Amazonia. In Vigia, Antônio Igo Palheta Soeiro, Paulo Cordeiro, and Seu Cebola (Ilson Pereira de Mello) were my hosts and accompanied me in my visits to Cacau, Ovos, and Santo Antônio do Tauapará. Thanks are due the Sociedade 5 de Agosto as well. In Santarém, Mary Jane Moreira Silva and Ayrton Pereira dos Santos graciously allowed me to visit the Arquivo do Foro. In Alenquer, Wildson Queiroz and Maria Ilka S. Cabral provided invaluable information about the city, and João Ubaldo Ribeiro, Áurea Nina, Guilhermo Antônio Martins de Araujo (Potyguara), and Marlucy Monteiro from the Toninho Notary granted me access to their personal and institutional archives. Thanks to Pacoval's Dona Cruzinha (Maria da Cruz de Assis), Dona Nezi, Zé Maria, and others who hosted me and had the patience to respond to my innumerable questions. In Óbidos, Dona Idaliana Marinho de Azevedo and Dona Anezia from the Museu Integrado de Óbidos, and Jorge Ary Ferreira of the Cartório do Segundo Ofício de Óbidos allowed me to use their impressive records. Not far from there, in Oriximiná, Maria de Souza, Hugo de Souza, and their family allowed me to stay with them as they shared some oral traditions with me. Many other members of Associação de Remanescentes de Quilombos do Município de Oriximiná or ARQMO also did so while I was there. Carlos Printes took me upriver along the Trombetas, where we visited Boa Vista, Jamary, Tapagem, and Abuí. Carlos also welcomed me into his home and into the community of Boa Vista, where I always felt at home. In Boa Vista we even watched the final game of a Champions League, which FC Barcelona won, of course.

Ana Lúcia Araújo, Manuel Barcia, Peter Beattie, David Blight, Anthony Bogues, Alex Borucki, Jerry Dávila, Marcela Echeverri, Anne Eller, Flávio Gomes, Mark Harris, Mary Karasch, Jane Landers, Hal Langfur, Jeff Lesser, Maria Helena Machado, Mary Ann Mahony, Jason McGraw, Alida Metcalf, Yuko Miki, Fabrício Prado, Tom Rogers, Heather Roller, Ted Rugemer, Lise Sedrez, John Soluri, Barbara Sommer, David Spatz, James Sweet, Barbara Weinstein, and two anonymous readers provided suggestions, advice, and encouragement for different parts or the whole of this book. Akin Ogundiran, Tanure Ojaide, Debbie Smith, Danielle Boaz, Dorothy Smith-Ruiz, Honoré Missihoun, Felix Germain, Jurgen Buchenau, Greg Mixon, John Cox, Erika Edwards, Carmen Soliz, Jill Massino, Eddy Souffrant, Sebastian Cobarrubias, and Maribel Casas also molded this study with their encouragement and their collegial support. Matthew Casey was absolutely instrumental to it. He is a colleague and a friend as generous as no other in the universe.

The University of Pittsburgh and its Center for Latin American Studies; the University of Central Oklahoma; Duke University and University of North Carolina (UNC) Chapel Hill's Consortium of Latin American Studies; University of North Carolina at Charlotte; Yale University's Gilder Lehrman Center for the Study of Slavery, Resistance, and Abolition; Brown University's Center for the Study of Slavery and Justice; and my very own Department of Africana Studies at UNC Charlotte provided the necessary funding to complete this book. I thank also Manuel Álvarez and Mia Storey for their work as research assistants at UNC Charlotte. Lyman Johnson, Elaine Maisner, and Michelle Witkowski also made this book a much better one.

To all my sponsors, readers, interviewees, guides, hosts, colleagues, assistants, friends, and relatives, I owe infinite gratitude. Release all sailors and fellow captains from any responsibility for errors in this ship's course—they are all attributable to its captain.

The People of the River

Introduction

Two Stories about Rivers, People, and Politics

In 1921, the state government of Pará, in Brazilian Amazonia, sent a special envoy to the municipality of Alenquer, near the confluence of the Tapajós and the Amazon Rivers, to supervise the privatization of lands containing Brazil nuts. In the course of his visit to Alenquer, civil engineer and legal expert Palma Muniz heard the grievances of poor smallholding cultivators who protested against the alienation of Brazil nuts from the public domain. Like rubber, the nuts were not cultivated, but collected from the forest. Unlike latex, however, the nuts had until very recently been seen as "belonging to the people," explained a local peasant; hence, the injustice of permitting propertied individuals to buy them from the government. But just after he embarked on a ship to proceed with the inspection of the Brazil nut areas upriver, a launch with forty people approached his vessel, its passengers cheering "for the governor . . . for Palma Muniz, for the blacks of Pacoval, and for the Curuá River in liberty." In the report that he would later write, Palma Muniz explained that the protagonists in this shocking incident were the "BLACKS OF PACOVAL," the descendants of a settlement of runaway slaves that had existed in the area at the time of slavery. "After denouncing some Alenquer merchants," Muniz related, the inhabitants of Pacoval demanded that the Curuá River became "a sort of patrimony of the blacks, so that only they could collect nuts [there]."[1]

Nearly seventy years later, in October 1990, the inhabitants of a black town from the municipality of Oriximiná, also not far from the confluence of the Amazon and the Tapajós, approved a manifesto addressed to the citizenry of the region rejecting the installation of a hydroelectric plant on the Trombetas River. "Our people cannot survive without Mother land. Without mother Water [sic]. We draw our sustenance from them," explained the statement they sent to the local and state media. Adopting a modern environmentalist tone, the manifesto continued to elaborate on how "Amazonia gives us LIFE. [From her] We receive fish, game, fruits, Brazil nuts, medicines, fibers, wood, shingles, mats, our homes' walls, [and] the manioc flour we eat." After suffering both legal and illegal mining activities in the area and the creation of a biosphere reserve behind the locals' back, "now ALCOA (USA) wants to implant yet another mining project on our lands." The central point in the manifesto was the rejection of that project: "We do not

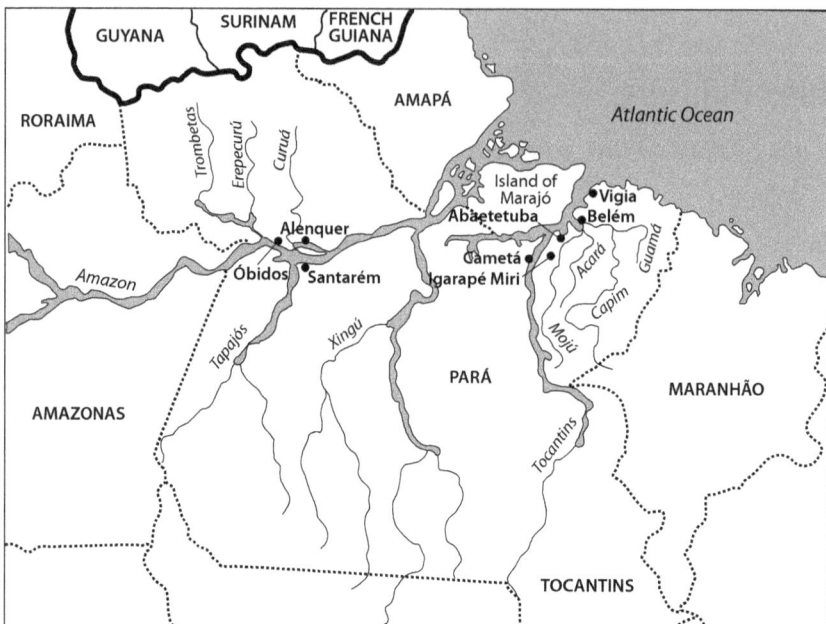

MAP 1 Selected cities and rivers of Amazonia, state of Pará, Brazil.

want either state-owned or private hydroelectric plants in our lands . . . . We want to be recognized as a people. We want . . . [the demarcation of] the forests and lakes that we have traditionally occupied as maroons [i.e., runaway slaves] for our community."[2]

These two stories are separated by nearly seventy years, but they share a number of ideas, images, and demands. Both portray villages inhabited by black farmers in the middle of Amazonia, a region usually seen as not belonging in the African diaspora. Both speak of rural Afro-Brazilians claiming publicly their right to enjoy free and unmediated access to forest resources, a claim they see as legitimated by the force of their agro-ecological traditions. And in both cases, we observe how groups of peasants make explicit reference to their African ancestry in a public statement, which sounds surprising in light of the emphasis on class identities in most studies about the rural populations of Brazil in the twentieth century. In these two discourses, the natural landscape is portrayed not only as vital to the black farmers' economy—it becomes an intrinsic part of their historical origins and of their political rights as Brazilian citizens as well. Observe, for example, their claims to maintain "the Curuá River in liberty" as "a sort of patrimony of the blacks" or their demand "to be recognized as a people."

The second of these stories comes from a period that we know relatively well. The Trombetas River peasants published this manifesto in 1990 because there was a chance the national government would intervene in their favor then. In 1988, Brazil had approved a new constitution that put the military dictatorship to an end, and Article 68 of its transitory dispositions stated that "definitive ownership will be recognized, and the respective title will be issued by the State, to those descendants [*remanescentes*] of the maroon communities occupying their lands."[3] Almost smuggled into the constitution by black movement activists and phrased as if it were only meant to benefit descendants of maroons (i.e., runaway slaves), Article 68 permitted the Trombetas River communities and hundreds of others throughout Brazil to obtain land deeds and official recognition from federal and state agencies during the 1990s and 2000s. As of early 2016, the Brazilian government recognizes the existence of 2,648 black rural communities; 1,533 of them have started the application for collective title to their lands, of which 207 have received the title already. Almost 16,000 families live in the two and a half million hectares of these 207 titled communities.[4]

Few people imagined in 1988 that Article 68 would make a big impact. The black activists who pushed for its inclusion knew that the constitutional delegates' only reference to settlements of runaway slaves, known as *quilombos* in Brazil, was to the famous quilombo of Palmares, a relic from seventeenth-century northeastern Brazil. Maroon descendant individuals, also known as *quilombolas*, seemed like a phenomenon belonging in the past, not existing in the present.[5] But after the constitution was passed in 1988, the concerted action of rural Afro-descendants, urban black activists, NGOs, scholars, and progressive legislators transformed the legal figure of the quilombo into a reparation for slavery, a settlement for Brazil's debt owed to African descendants since slavery's abolition in 1888. Thus, in the present any "self-designated ethno-racial groups with their own historical trajectory, specific territorial relations, and a presumed black ancestry related to the historical oppression they have suffered" may apply to be recognized as quilombo descendants, becoming eligible to receive a collective land deed and other social welfare programs after a long bureaucratic process.[6]

The black communities, also called black lands (*terras de preto*), went from unknown to big news almost overnight. During the 1990s they participated in the Brazilian government's gradual awakening to the existence of racism, culminating in the creation of the Brasil Quilombola program in 2004. A comprehensive agenda coordinating housing, land titling, and social programs,

Brasil Quilombola has been one of the main vehicles for the gradual empowering of the communities ever since. Since the 1990s, the communities also enjoyed a sustained presence in the media, both Brazilian and international, including the *New York Times*, NPR, BBC, *The Economist*, *Huffington Post*, and Al-Jazeera, among other international media.[7] In the academic world they have become relevant in discussions about anti-racist policies not only in Brazil but also in the rest of Latin America.[8]

Most scholars writing about the quilombolas' history have emphasized the constructed nature of their racial identity. U.S. anthropologist Jan French, for example, highlights in her *Legalizing Identities* "how the invocation of laws can inspire ethnoracial identity formation," because those laws provide "structures for self-identification, mobilization, and social justice."[9] Others have argued that "taking on [an] identity as a quilombola is just one strategy among many for reducing their economic marginalization," a strategy used by urban black movements, progressive Catholic Church activists, and some anthropologists to "introduce" new "ethnic categories" of identification and political action that did not exist before.[10] In sum, according to a number of scholars, the 1988 constitutional clause underlies the present-day process of black peasants' identity-building and political mobilization.

In contrast, this book argues that, despite the changes set in motion by Article 68, a number of discourses and practices inherited from past generations continue to be a source of identity for the black peasants of Brazil. Dating from the era of slavery, such traditions have subsisted embedded in the landscape in the form of agro-ecological strategies, political discourses, and even relationships with political and economic elites. They have not resulted in hard ethnic boundaries sharply separating black peasants from other rural Brazilians, but rather have been part of a flexible toolbox of strategies and narratives inscribed on the landscape and used in moments of conflict over land, labor, and citizenship. While the relationships between black rural communities and the natural world have been largely overlooked as a vehicle for the maintenance of an Afro-Brazilian identity, this book is devoted to unearthing their existence, interrogating their relevance, and putting them in dialogue with the broader history of slavery and its legacies in post-emancipation Brazil. The rhetorical figure of the "people of the river," which epitomizes those relationships, titles this book.

Nature, Community, and the Transition to Freedom

In the last few decades the historiography discussing the legacies of slavery among rural Afro-Brazilians has flourished, reaching a high degree of analytical depth and conceptual sophistication. Anthropologists and historians, for example, have carried out studies of present-day black towns, and from there they have traced their past. Works such as Antonio Candido's *Os Parceiros do Rio Bonito*, Harry Hutchinson's *Village and Plantation Life in Northeastern Brazil*, or the more recent *Cafundó: Africa in Brazil*, by Peter Fry and Carlos Vogt, pioneered this approach in the mid- to late twentieth century. A large number of monographs on black communities followed suit during the 1990s and 2000s as Article 68 led hundreds of communities to seek official recognition.[11]

This book partially follows this approach by focusing on the trajectory of maroon descendants in frontier Pará. The maroons, known as *mocambos* in Amazonia, were stable settlements of runaway slaves that existed on the peripheries of slave societies—and occasionally in their interstices—all over the Atlantic world. These settlements evolved over time and space, from early colonial communities resembling African political and religious entities, to semi-tolerated settlements pushing to accelerate the arrival of abolition.[12] The main paradox about the quilombos, as they are usually called in the rest of Brazil, is also their most recognizable characteristic: their attempt to prevent reenslavement through the use of defensive mechanisms, from being in inaccessible locations to creating networks of informants, while at the same time maintaining contact with the broader society. Only thus could they have access to trade, marriage partners, or even peace negotiations with colonial and national governments. Keeping this focus on relationships with the broader world, in this study I illuminate how maroons became free peasants through a process of negotiation that started under slavery and that involved missionaries, merchants, political patrons, and, above all, Brazil nuts.[13]

However, *The People of the River* also seeks to enrich the history of marronage by adding an environmental perspective. Because the natural world around the maroon communities was instrumental to their project of freedom and autonomy, I discuss the maroons' history by interrogating the relationships between the natural and cultural landscapes they inhabited. Landscapes are "the symbolic environment[s] created by a human act of conferring meaning on nature and the environment"; in other words, they are the sum of a physical place plus the narratives people formulate about it. Because a landscape is simultaneously "what we see and how we see it," it is a very useful concept for illuminating the connections between the natural world and historical processes of material and cultural change. By treating black Amazonia's landscapes as a "message which needs to be decoded" and by disentangling how they came to be "spaces of personal and social identity," I seek to open a new window into the transition from slavery to peasantry in Brazil.[14]

While some of Amazonia's black towns descended from maroon settlements, others originated in the slave quarters. *The People of the River* also locates the roots of Brazil's black peasantry in the learning of an independent livelihood by those living in the *senzalas* or slave quarters. Here the analysis interrogates "the spaces of autonomy" conquered by slaves in their workplaces, such as "a rhythm of work decided by the group, an independent social organization, an incipient subsistence production in provision grounds[,] and a monetary micro-economy."[15] The existence of these hard-won victories explains why, even in the highly capitalized coffee- and sugar-producing areas of Brazil we find examples of "black peasant micro-communities"; that is, groups of freed people who despite the abolition of slavery in 1888, or maybe because of it, "stay[ed] on the properties" where they had worked as slaves.[16] In sum, in this study the transition from slavery to freedom is to be found both inside and beyond the plantation.

Of special importance in this process is what American historians have called the internal economy of slavery—the activities the enslaved carried out for themselves—with or without their masters' permission. Based on the "obtention of their own house and a provision ground," which usually implied the right to work "for themselves," the slaves' internal economy could attain a considerable complexity, as evinced by the multiple crops or the varied and singular activities that allowed the enslaved to obtain small amounts of cash in the United States, Brazil, Jamaica, or Cuba.[17] In the long term, the slaves' economy became a "right" and permitted them to attain a certain "family stability," therefore generating some "limits" to the masters' near-absolute power.[18]

I found it necessary to adapt this concept of the internal economy to the particularities of Amazonia. As in other plantation regions, in this tropical forest the cultivation of a *roça* or manioc ground was often at the center of this economic system. However, the extraction of forest items (aka extractivism), the hunting and raising of different species, from turtles to manatees to livestock, and the relations with itinerant canoe peddlers known as *regatões* could occupy much more of the slaves' time than agriculture. In addition, mastery over local environments was more rewarding in Amazonia than in other Brazilian regions. Specializing in forest resource management, canoe navigation, exploration, tourist guiding, or accounting in a trade store, for example, frequently represented viable paths to new sources of income and even to stable jobs. Finally, the tropical forest of Amazonia had numerous similarities to its counterpart in the Congo-Angola region, which facilitated transfers of knowledge and technology between both shores of the equatorial Atlantic.[19]

For all these reasons, I replaced the concept of the slaves' internal economy with that of the slaves' *parallel* economy: the productive and commercial activities they carried out in order to participate in Amazonia's rapidly expanding commercial networks during the second half of the nineteenth century. This parallel economy was not subject to the spatial limits of the plantation, but instead nourished, and fed off, the broader export trade taking place in these decades, fueled of course by rubber. While in most plantation regions the internal economy of slavery often consisted of a series of contained farming activities sometimes accompanied by trading, in Amazonia the slaves' parallel economy was just an illicit branch of an enormous commercial network, as shown by the slaveowners' angry lambasting of itinerant trade during the last decades of slavery.[20]

The erection of a parallel economy could only take place after slaves and maroons explored Amazonian landscapes and came to understand their seasonal rhythms, their marketable products, their possibilities for agriculture, and their potential to sustain families and communities. I call this process of acquaintance with the opportunities and constraints of local environments "environmental creolization." Usually, the concept of creolization designates the cultural and demographic adaptation of enslaved Africans to life in the Americas, involving processes of cultural mixture and change, occasional re-Africanizations, and the flexible maintenance of numerous African cultural referents.[21] However, I apply the concept not only to the initial arrival of Africans to the region but also to their broader strategies of social and economic advancement throughout the entire era of slavery. Just as they learned a new language, became Brazilian by birth, or came to understand the interstices of Brazilian law to gain autonomy or even freedom, slaves and maroons also acquired knowledge of the natural world with the intent of becoming forest collectors, hunters, rangers, or farmers. With the concept of environmental creolization, then, I intend to capture the ideas, practices, strategies, and discourses that both African and Afro-Brazilian slaves learned not only as they adapted to the New World, but as they sought to become free peasants as well.

Some scholars have already focused on the importance of natural landscapes for the history of African slaves in the New World. Geographer Judith Carney has investigated, for example, how the "distinct cultivation system" of rice that came from Upper Guinea to the Americas in the eighteenth century permitted some slaves to work as technicians in erecting hydraulic infrastructures near rivers and swamps and provided the free population of color with a more diverse diet. Enslaved people coming from Kongo to the South Carolina

Lowcountry also impregnated water landscapes with simbi, water spirits that guarded the natural world, regulated the passage between life and death, and allowed the captives to "anchor communities of the living" in the region. Others scholars have interrogated the "creole ecologies" that emerged in the colonial Americas or have established how natural landscapes offered "highway and sanctuary" for runaways, providing "resources" to "consolidate the [slave] community" for those living in plantations and becoming a resource with the potential to change the legal status of large numbers of captives.[22] Finally, they have also established that, just as the Atlantic plantation complex ruthlessly consumed human lives to produce sugar and other crops, it wasted natural resources just as recklessly.[23]

A focus on the process of environmental creolization has the potential to generate a new perspective on the history of slavery. If we consider relationships with the natural world as one more battlefield for contests over autonomy and dignity, then the slaves' discourses and entanglements with natural landscapes become a window into the "histories, experiences, and interpretations" behind their "visions or definitions of freedom and captivity."[24] They enable us to encounter strategies of autonomy whose significance has been hiding in plain sight, such as that of the enslaved individuals analyzed in chapter three whose forest-related skills, whether learned in Amazonia or partially brought in from Africa, helped them become free of their legal and social status of slaves. If the essence of the plantation complex was bringing an "exotic work force" to the New World, then the process of environmental creolization experienced by the African newcomers represented an attack on, and ultimately the defeat of, the rationale behind planter capitalism.[25]

The first part of this book, in sum, is a social and environmental analysis of how the enslaved became peasants. Chapter 1 discusses plantation slavery in Pará in the decades after the Cabanagem revolt, an enormous anti-Portuguese rebellion that shook post-independence Amazonia between 1835 and 1840. It shows how slaveholding planters of tropical crops successfully adapted their productive strategies to the Amazonian environment in a context of destruction and lack of capital to rebuild their estates. It also discusses how the revolt reinforced the demographic creolization of the slave labor force by severing the region from the slave trade, which balanced the sexual ratio of the captive population. Chapter 2 and 3 offer a microsocial perspective on the process of environmental creolization. The former uses the oral myth of the Big Snake to analyze how maroons in the Lower Amazon merged fragments of West African and Indigenous mythology to narrate their experiences and how instrumental their mastery over local environments was in resisting the slaveholders' might. Chapter 3 takes us to the *senzalas* or slave quarters by visualizing how that same mastery could alter the terms of daily life under slavery. Taking the example of Larry, an English-speaking slave who once told Charles Darwin's colleague Henry Bates that he "did not want his freedom because he was not a fool," this chapter explores the relation between the learning of a rural living in Amazonia and the subversion of slavery, including the side effects.

Afro-Descendant Identities and Post-Emancipation Claims to Citizenship

There is something else about the two stories from 1921 and 1990 at the beginning of this introduction that probably caught your eye. Both of the black communities quoted there expressed their political goals through idioms of belonging to the natural landscape: the "free" Curuá River and the "mother Land, Mother water" that the inhabitants of the Trombetas River invoked. These references are far from trivial, and in the 1990 case do not arise exclusively from the influence of late twentieth-century environmentalism—although such influence existed, of course. Rather, these references to the natural world speak of the centrality of discourses and practices nested in nature to the maintainance and reconfiguration of an Afro-descendant identity in the post-emancipation decades. Both communities explicitly invoke agroecological models deeply entwined with their past as Afro-descendants and defend them vis-à-vis state institutions, even in a period like the 1920s, when their black identities were supposed to be largely irrelevant.

Indeed, the existence of a black peasantry—a group of smallholding free cultivators with an awareness of being Afro-descendants—has traditionally received scant attention in Brazilian historiography. Pioneering intellectuals portrayed the passage to freedom as the incorporation of rural Afro-Brazilians into a mixed-race peasantry. In 1918, Afro-Bahian intellectual Manuel Querino praised the "hard working" African slave by giving him "a place of honor as a factor of Brazilian civilization," but then quickly merged him into the "mestiço" or mixed-race population after emancipation. In 1933, the famous sociologist Gilberto Freyre also understood Brazil's free rural population as "mulatto in composition, or a mixture . . . of three races," and suffering actually worse living conditions than it had experienced under slavery, because it was "deprived of the patriarchal assistance of their former masters and of the diet of the slave quarters." "Materially and psychologically ravaged" by slavery, the no less prominent sociologist and politician Florestan Fernandes argued in 1964, after abolition the ex-slaves "lacked the means to assert themselves as a separate social group," because they could not form an "independent cultural horizon."[26]

Modern historians have argued that there were significant obstacles preventing the generation of a shared identity among the enslaved, and therefore of generating a black, or Afro-descendant, post-abolition peasantry. Bahian scholar Luís Nicolau Parés, for example, emphasizes the existence of a "stratification based on both color and origin" among the enslaved in eighteenth- and nineteenth-century Bahia, "with mixed-race individuals on top, Brazilian-born slaves in the middle, and Africans at the bottom." Stuart Schwartz, Manolo Florentino, João José Reis, and James Sweet note the presence of similar obstacles in both rural and urban spaces. Sheila de Castro Faria explains that, because of those obstacles, "few regions [in Brazil] could have the conditions to create a singular slave community" encompassing all the captive population in that region. Hebe Mattos has a similar perspective: "the family and the slave community did not become the matrix of a black identity stemming from captivity, but [conceived] in parallel to freedom."[27]

But others have identified some social structures that were capable of sustaining a black identity in the post-abolition decades. "The bonds of community woven through kinship . . . were fundamental for the freedpeople to rebuild their lives in freedom," argues Walter Fraga Filho when describing the sugar-producing area of the Bahian *Recôncavo*. Reinforcing those bonds was "a memory, transmitted from parents to children," underpinning a "unique personality" as black peasants that could be modified "according to the context."[28] Black rural communities could be formed by "maroons" or by "freedmen and sharecroppers," because they frequently "interbred" with each other once slavery ended, as Flávio Gomes, Maria Helena Machado, and others emphasize. In sum, we should pay heed to Faria when she argues, "In Brazil slavery formed slave communities, in plural"; that is, a black peasantry with an identity strongly shaped by local components.[29]

Natural landscapes constitute an ideal window through which to observe this process. They became "part of the shared symbols and beliefs" of the black peasants, just as the elites used natural tropes in an analogous but opposed way to inscribe "their place atop the economic, racial, and social structure."[30] Thus, interrogating "the manifold ways class [and racial] identities have been forged . . . in the natural and built environments" allows us to discern how sugar elites in early twentieth-century northeastern Brazil "embedded in the sugar fields around them a history of their own naturalized power," or how in present-day Virginia's Poplar Forest plantation "racialized tensions embedded in the physical and social landscapes" still manifest "the complex social relationships between slaveholders and enslaved, landowners and workers, and Blacks and Whites."[31] This study argues that natural landscapes were instrumental to both the preservation and the reconfiguration of a black peasant identity in the decades after emancipation.

The second part of this book, then, interrogates the role played by ideas of place and nature in conflicts over land, labor, and politics—the realms where black peasants mobilized their collective identities to claim for citizenship. While in the first chapters of the book I discussed separately how plantation slaves and maroons made it into the ranks of the peasantry, here I gradually erase the differences between both groups and end up treating them as a single black peasantry—albeit one with significant local and regional variations. Chapter 4 traces the conflicts between maroon descendants and merchants over the extraction of Brazil nuts. It emphasizes the role of *mateiros* and other specialists in forest activities in the process of dispossession that took place in the Lower Amazon in the 1910s and 1920s. Chapter 5 investigates the expression "citizens of [the island of] Tauapará," the catchphrase an Afro-descendant elderly man used when remembering how a group of freedmen won a land dispute with a cattle rancher in 1944. The expression bears witness, I contend, to how the black peasants constructed a discourse of belonging in a dialogue with the law and with other sectors of Brazilian society. Finally, chapter 6 crowns the book by disentangling the cycle of protests that shook the Lower Amazon in the 1920s, when black peasants protested the privatization of lands containing Brazil nuts. That cycle contributed to the growing popular discontent that brought President Getúlio Vargas to power in 1930.

Introducing African Slavery in Amazonia: From Pombal to Cabanagem

The manifesto from 1990 discussed earlier surprised the Brazilian public for another reason. Traditionally, Amazonia has been seen as a region populated mostly by Indians, Portuguese colonists, and the mixed-race free peasants known as *caboclos*. In 1900, only twelve years after the abolition of slavery in Brazil, Paraense writer José Veríssimo explained that among all Brazilian regions Amazonia was "among the less populated by blacks" and that "today it is extremely rare to find Africans in the two provinces [Pará and Amazonas] outside of their capitals." Historians and economists reiterated this portrayal during the twentieth century. "Abandoned by the Portuguese government" during the colonial period, explained Brazilian economist Celso Furtado, the captaincy of Pará "lived exclusively on the forest-extractive economy organized by the Jesuit fathers and based on the exploitation of Indian manpower." "By and large," added the equally prominent Caio Prado Jr., "the region continued to depend almost entirely on the gathering of indigenous forest commodities." American historians followed suit: in the 1970s Colin MacLachlan assessed the Portuguese Crown's attempts to import enslaved Africans to Amazonia as having simply "failed," based on the fact that by 1800 the 18,944 slaves brought to Pará represented only 23 percent of its population, in comparison to neighboring Maranhão's 40 percent. History books still describe the "peripheral North" as a region where "compulsive indigenous labor predominated," erasing from the region's historical record activities other than the extraction of backland drugs.[32]

However, African slavery did represent a key part of Amazonia's economy. There were initiatives to bring enslaved Africans to Amazonia ever since the Portuguese colonized it in the seventeenth century. Barely decades after the Portuguese settled in São Luís and Belém in 1615, the Crown became directly involved in fostering commercial agriculture by donating lands to colonists who cultivated cash crops and by signing contracts with private agents to introduce African slaves. At least eleven contracts were signed between 1680 and 1702, eventually bringing approximately 2,000 slaves to the region.[33] This was a small number in comparison to the approximately 78,000 slaves whom the other Brazilian ports received in the same years, although significant given the small population of Paraense cities in the period.[34] In the end, this number of slaves was clearly insufficient to spur the emergence of a plantation economy.

This would have to wait until Sebastião José de Carvalho e Melo, count of Oeiras and marquis of Pombal, became the empire's political strongman under the new king José I in 1750. At a time when the French and British Empires developed rich plantation colonies in the Caribbean, Pombal was concerned with not losing ground to them—in both economic and military terms. The marquis devised a plan that included reorganizing Amazonia's economy, restructuring the management of Indian labor, promoting extractivism, and building a plantation economy.[35] The main tool he devised to do so was a joint-stock monopoly trading company to import enslaved Africans. Financed through private capital and favored by the Crown with numerous tax and legal privileges, the Companhia Geral de Comércio do Grão-Pará e Maranhão began introducing African slaves to São Luís and Belém in 1755.[36] The company would make it possible, Pombal and his brother Francisco Xavier de Mendonça Furtado thought, to "cultivate the land and multiply its fruits until they become abundant" or, in other words, to fully develop Amazonia's dormant agricultural and commercial might.[37]

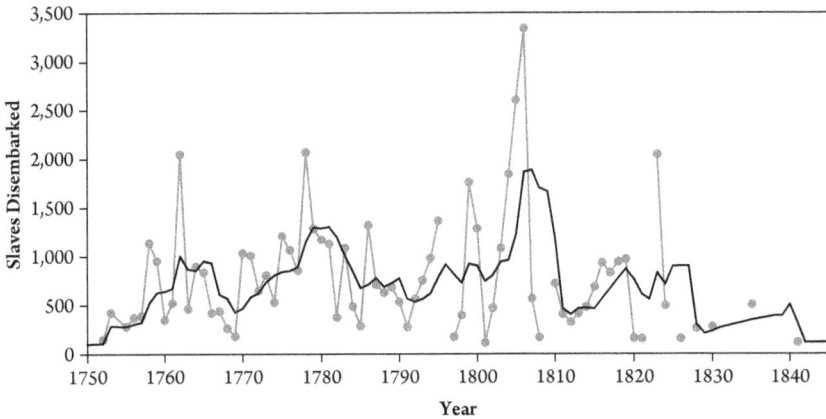

GRAPH 1 Slaves disembarking in Belém, 1750–1841, with five-year moving average. Source: Highest annual values from either the Trans-Atlantic Slave Trade Database, www.slavevoyages.org (accessed May 5, 2014), or Bezerra Neto, *Escravidão negra no Grão-Pará*, 208–16.

During its twenty-three years of existence, the Companhia Geral invigorated the commercial economy of the state of Grão-Pará and Maranhão, increasing exports of items already cultivated, such as wild cacao and other spices, and introducing new ones like rice or cotton. More importantly, it also created new and enduring links between Amazonia and the Atlantic coast of Africa. During the company's first three years of operation, almost 6,000 slaves were sold to local planters and merchants in Belém, a shocking 50 percent increase over the number sold in the preceding half-century. Between 1756 and 1778, the company imported an average of almost 844 enslaved Africans per annum, for a total of 18,563 in the span of twenty-three years, as shown on graph 1. Slave imports remained high between 1778 and 1800, with another 17,323 captives entering the port of Belém at an annual average of approximately 825. The state economy had expanded thanks to the company's activities, yielding enough benefits to finance the regular purchase of enslaved Africans.

Approximately two-thirds of the slaves entering Belém (12,375 of the total of 18,563) came from Upper Guinea during the company's existence; the other third came from West Central Africa and São Thomé.[38] But while the neighboring captaincy of Maranhão would retain an ethnic majority of slaves from Upper Guinea until 1815, the consolidation of the Portuguese presence in Angolan ports around 1750 led to a gradual increase of slaves from that region in the 1780s. While Maranhão's slave communities maintained a marked Guinean identity through interethnic marriage, rice consumption, and a common core of religious beliefs, the cultural constellations that African slaves brought to Pará tended to be more ethnically mixed, and eventually much more West Central African in character.[39]

The importation of African slaves to the port of Belém peaked between 1800 and 1810; in this decade 10,927 slaves, or almost 1,100 every year, disembarked in the city. While the Napoleonic Wars in Europe and the arrival of the Portuguese royal family in Brazil in 1808 clearly disrupted the traffic after 1808, the middle years of the first decade of the nineteenth century witnessed the largest arrivals of slaves ever recorded, reaching a startling 3,339 captives entering Belém in 1806—almost the same amount of slaves imported that year to Recife, one of the main slave ports in Brazil.[40]

In the following two decades there was a gradual decrease in the import of African captives; the slave trade was to disappear completely by 1830. Belém's commercial activities were seriously damaged when the Portuguese royal family and court moved to Rio de Janeiro in 1808, leaving continental Portugal in the hands of Napoleon's armies. Since the slave trade to Amazonia relied on the connection with Lisbon merchants—much more so than in the State of Brazil[41]—the disruptions of the Portuguese slavers' activities and of commercial activities in Lisbon represented a serious challenge for Belém's merchants. The falling prices of cacao in world markets during the 1810s, and the Anglo-Portuguese occupation of French Guiana between 1809 and 1817, directed from Belém, also probably siphoned off capital from the city's slaveowners for the purchase of slaves.[42] Between 1810 and 1820 only an average of 617 slaves entered Pará annually, returning almost to pre-company levels.

By the 1820s, Grão Pará was feeling the political instability caused by the conditions that led to the Cabanagem revolt. By practically eliminating the slave trade to Belém and Maranhão and by shaking the foundations of the Paraense economy, this revolt in 1835 brought the expansion of slavery to an end. For about eighty years the main ports of the state of Grão-Pará and Maranhão—Belém and São Luís—had participated vigorously in the Atlantic slave trade, which had brought 59,184 enslaved Africans to the state and later the province of Grão-Pará. They were the last Brazilian port cities to enter this inhuman trade and the first to leave it, but for some decades they constituted a vigorous branch of it that extended as far as Lisbon.[43] Despite the short-lived nature of the slave trade in Amazonia, Africans and their descendants became part of the region's history, shaping its course in the decades and centuries to come.

CHAPTER ONE

After the Reign of Terror

Slavery and the Economy of
Post-Cabanagem Pará, 1835–c. 1870

Laboring at the São Miguel da Cachoeira property, on the Guamá River, must have been an exhausting activity for the 52-year-old slave Efigênia. The work done during the last two months of 1840 was particularly backbreaking. João, Joaquim, and the other male slaves had cleared some fields to plant rice before the rains started in December, working from dawn to dusk cutting down small trees, removing rocks, and uprooting weeds. Efigênia and the other slave women planted, weeded, milled, and winnowed the rice, tended to other food crops, and cooked. Her only respite probably came when she got home at night, perhaps in time to see her two-month-old grandson Lázaro and forget momentarily the harshness and exhaustion of work. The little baby may have encouraged Efigênia to think that there was a future ahead for her family, as did the other babies born that year from slaves, such as Simplício and Francisco, and the other nine toddlers who lived in the plantation. They were not yet considered field hands as they were not even 10 years old, but that would change soon.

Living with so many children and elderly individuals, however, was a mixed blessing for old enslaved workers like Efigênia. Adults like her Mozambican husband Joaquim or the Congolese slave João were getting too old to work in the fields and carry heavy loads. Pedro, of Angolan origin like Efigênia, worked clearing the fields, planting, and transporting the sacks of rice, which cost him a hernia at age 38. Germana probably worked as much as she could at her age, and so did Efigênia's daughter Marcelina; yet both had given birth to two babies recently, and since they carried their newborns around, their capacity to work inevitably decreased. Ten of Efigênia's fellow slaves, almost half of the entire crew of twenty-one, were 10 years old or younger, and another eight were older than 30—with those between 40 and 50 being the largest group. In sum, of the twenty-one enslaved individuals living in Bernardo José Paes Júnior's rice plantation in 1840, only a handful were adults in prime working age. This is what this plantation was, Efigênia may have thought that night in late 1840: a place where mostly elders and children worked.

Efigênia's story epitomizes the characteristics of post-Cabanagem Amazonia's slavery. It is a prime example of how Paraense planters adapted the production of tropical crops for export to the region's agro-environmental conditions. Because most Amazonian soils are poor in nutrients, producing cash crops such as cacao, rice, or even sugar demanded agricultural techniques different from those employed in other regions of Brazil. This chapter examines those techniques used in cacao and sugar plantations. Cacao is a tree native to Amazonia whose fruit was exported since the colonial period; cacao producers relied on the river tides for soil fertilization and combined its cultivation with the exploitation of wild groves. Sugar producers adopted a similar approach, using tidal energy as a natural fertilizer and as a source of energy to propel their mills. Amazonia was able to successfully develop a slave economy that produced tropical crops for markets overseas.

The demographic profile of Efigênia's rice plantation in 1840 also reflects the impact on Paraense plantations of the 1835 Cabanagem revolt, a major social and political uprising against Portuguese elites. Most observers described the revolt as "a reign of terror" waged by popular, nonwhite groups, when "the plantations were burned down [and] the slaves and livestock massacred."[1] Nonetheless, its impact on the region's slaveholding properties is more complex than previously thought, as this chapter shows. While the loss of slaves in prime working age naturally affected the output of local plantations, the Cabanagem revolt also reinforced the trend toward a sexually and ethnically balanced slave population inherited from the pre-Cabanagem era, which was key to the natural reproduction of the enslaved population in later decades. Despite the damage caused by the rebellion, then, the post-Cabanagem slave demography of Paraense plantations turned out to be favorable to the reconstruction of the state's plantation sector and to the natural reproduction of its enslaved laborers as well.

Ultimately, the slave sector of the economy represented a key activity—perhaps the most important one—in the economic recovery of Pará between 1835 and approximately 1870. While rubber is still seen as the undisputed king of Amazonia's economy in the second half of the nineteenth century, in reality the export of slave-produced plantation crops was among the main forces behind the province's post-Cabanagem growth.[2] Cacao, sugar, rice, manioc flour, meat, and hides kept Pará's economy alive and kicking, propelling its planter elite to the highest levels of Imperial power. Wealthy planters like Antônio de Lacerda Chermont, viscount of Ararí, or Antônio Gonçalves Nunes, baron of Igarapé-Miri, occupied the most important seats in Pará's institutions until the rubber barons took over. But let us start this story in a sweet and tasty way—by talking about cacao and sugar.

Cacao and Sugar Cultivation in Nineteenth-Century Amazonia: Agro-Environmental Perspectives

Cacao was Amazonia's main export commodity since the colonial period. During the eighteenth century, Indian laborers under governmental and private supervision collected the wild cacao or *cacau bravo* in canoe expeditions to the interior of the region that lasted for months.[3] Gradually, however, the production of cultivated cacao grew, and by the 1800s it probably surpassed the wild form. Between 1800 and 1809, Amazonian plantations supplied more than one-third of all the cacao entering Europe and, in the following decade, about one-quarter of it.[4] By then, cacao plantations had substantially altered the natural and social landscapes of the Lower Tocantins and the Lower Amazon regions. In 1827, British explorer Henry Lister Maw chronicled how near Óbidos "the cocoa plantations . . . extended for miles along the banks of the river," encroaching on the free peasants' properties and contributing to the unrest that led to the Cabanagem revolt in 1835. Ten years later the naturalist Henry Walter Bates noted how "most of Óbidos townsfolk are owners of cacao plantations."[5] There are two main varieties of cacao: *forastero* and *criollo*. During the nineteenth century the two varieties of forastero known as *amelonado* (or *comum*), and Pará (or *calabacillo*) predominated in Amazonia. The *criollo* variety, originally from Central America, was also cultivated in the region, although probably to a lesser extent.[6]

Between 1847 and 1869, on average about 3,350 tons of cacao were exported annually from Pará—surpassing rubber in the export trade. Despite the latex boom in the last decades of the nineteenth century, or perhaps because of the commercial networks it created, cacao exports increased to more than 4,700 tons of cacao annually in the 1880s; between 1889 and 1918 they slowly declined, but still reached approximately 2,700 tons per year.[7] The rubber boom, in other words, did not put cacao exports to an end. What probably caused the decrease in cacao exports in the 1890s were the cacao boom in the northeastern province of Bahia and the nationwide abolition of slavery in 1888.[8]

It is logical that cacao maintained such a preeminent role in the nineteenth century. Both in its wild and in its cultivated form, the *Theobroma cacao* tree demands two elements in order to have a generous yield: a "tall and vigorous" vegetal canopy above it because it is "a member of the lower story of the rain

FIGURE 1 A cacao plantation in Santarém, 1874. Smith, *Brazil, the Amazons and the Coast,* 260. Courtesy of North Carolina State University Libraries.

forest," and a soil "rich in humus, [and] cool."[9] The first condition was well known by all Amazonian inhabitants: cacao was considered an understory crop since the colonial period. As per the second requirement, local cultivators had two methods to fertilize the soil: (1) planting other fruit trees such as bananas around the *cacaoeiros,* whose falling leaves eventually formed a thick layer of humus around the cacao trees, and (2) cultivating the cacaoeiros in the high floodlands or *várzeas altas,* strips of terrain inundated twice annually for a short period due to a combination of the Amazon's seasonal and daily high tides.[10]

The Theobroma trees grew from stem to six or seven meters tall in approximately three years; at that point they each started yielding between fifteen and thirty pods, each containing a few dozen seeds, that is, the beans.[11] The pods were harvested twice a year and the seeds left to ferment for three or four days in large boxes covered with banana tree leaves. After the beans hardened and acquired their characteristic scent, they were spread on wooden platforms and left to dry, which would take another four to five days—and twice that time in the rainy season. By then the beans had become a fragrant and attractive brown-colored, chocolatey fruit and were ready to be shipped abroad, where chocolate and other cacao-based products were made. Certain byproducts of cacao, such as the opened pods, could be used to brew a drink known as "wine," or to fertilize the soil around the trees.[12]

The canopy that cacao trees needed was often obtained by growing them under crops like manioc, castor bean plants, or fruit-bearing trees such as bananas, *urucurí* palms (*Attalea phalerata*), or *bacurizeiros* (*Platonia insignis*). In 1852, American explorer and U.S. Navy lieutenant William Herndon observed during his stay in Óbidos how "plantains, Indian corn, or anything of quick growth, are planted between the rows" of cacao trees, and Henry Walter Bates noted how cacao producers "might plant orchards of the choicest fruit-trees around their houses."[13] A staple well fit for both the natural and the social conditions of Amazonian landscapes, cacao continued to be ubiquitous during the nineteenth century in agricultural properties both large and small, as provincial president Francisco de Sá e Benevides signaled in 1875.[14]

In fact, small- and medium-sized planters often combined cacao production with other crops in the post-Cabanagem period. Engaging in a seasonal combination of agricultural and collection activities, these "portfolio cacao planters" represented a heterogenous group in terms of wealth and property, although slave crews tended to be small or medium sized. They mostly ranged from one or two individuals to ten or twelve, with children on average making up at least one-third of the crews.[15] Their cacao groves generally comprised between 2,000 and 6,000 cacao trees, but smaller groves of 200–500 trees were also common.[16] Produced in conjunction with a wide range of agricultural and other items, such as manioc, tropical fruits, coffee, annatto, cotton, or cattle, cacao was not necessarily the primary source of income for these farmers.[17]

Above these diversified producers there was a group of slaveowners with larger estates and fortunes that rested primarily on cacao—that is, cacao planters *tout court*. Most had slave crews of ten to thirty individuals, although small ones of five or six slaves also appear among their postmortem inventories between 1850 and 1880.[18] Instead of holding large fields of Theobroma trees, these cacao planters exploited a collection of large, medium, and small groves located either in different areas of the same property or in separate lots that could be up to several hours distance from each other. José Antonio Lourinho, a cacao planter from Igarapé Miri who passed away in 1856, had fourteen slaves working in three different cacao groves spread along the Pindobal River with 4,487, 1,720, and 8,000 trees, respectively. The following year, Maria Ritta Corrêa de Miranda bequeathed at least four different *cacoais* or cacao groves to her descendants: one with 1,300 trees at the Igarapé Borges, another with 3,000 trees at the Igarapé Cacoal, one more with 800 trees at the Manoel Ferreira Igarapé, and a last one with 1,500 more trees in her São José site.[19] In contrast to the smallholding farmers discussed earlier, the cacao planters tended to have larger slave crews and dedicated themselves almost exclusively to the mono-cultivation of cacao.

In his will from August 15, 1853, cacao planter Máximo de Miranda Portugal may have provided one of the reasons why cacao planters relied on collections of separate groves: his estate contained "some cacao trees purchased and others produced by nature."[20] In other words, in the post-Cabanagem period the planters cultivated the staple, but if groves of *cacau bravo* were available, the cacao producers would not hesitate to exploit them. Further contributing to the fragmentation of cacao estates was the fact that Brazilian inheritance law established that the heirs of a deceased person always received two-thirds of his or her estate divided into equal parts, an impediment to keeping large properties intact. Any existing debts at the time of the demise also demanded repayment, which led to the sale of parts of the estate and therefore to a further fragmentation of landed properties—and debt to commercial houses was certainly a structural feature of Pará's economy in the second half of the nineteenth century.[21] Nature, debt, and inheritance law, then, forced the post-Cabanagem cacao planters of Pará to rely on multiple groves to produce sizable amounts of the precious chocolatey beans.

A third ideal type of cacao producers comprised large landowners who planted the crop in large properties devoted primarily to other activities, such as sugar plantations or livestock farms.[22] For them cacao was simply one more cash crop, used to diversify their sources of income. Sugar planters, for example, found that the production of cacao was fully compatible with the cultivation of sugarcane: both crops thrived in the high floodlands of the Amazon tributaries near Belém: the Guamá, Acará, Mojú, Capim, and lower Tocantins Rivers. Due to the proximity of the Atlantic Ocean, the high floodlands or *várzea alta* are covered by the daily high tide during four to six weeks of the spring season, and sometimes for another two weeks around September. Able to withstand waterlogging for short periods, the cacao trees and the sugarcane benefit from the nutrients deposited on the soil during this annual cycle.[23] In a visit to Pará in 1873–1874, geologist Charles B. Brown and civil engineer William Lidstone described how "mandioca (cassava), maize, and sugar-cane frequently replace the cacoa [*sic*] trees" in the "narrow elevated strip[s]" of land "just above the level of the water." By the mid-twentieth century, sugarcane and cacao were still planted in the várzea alta of the rivers near the Amazon mouth.[24]

In addition to being planted in the same types of land, cacao and sugar were compatible for other reasons. Sugarcane was usually planted by May or June, so that it was at least three to four months old when the rainy season came in September, just before the second harvest of cacao took place.[25] Both plants were compatible with a number of food crops as they grew: rice, corn, beans, squashes, and short-time manioc in the case of sugarcane, and banana, rubber trees, fruit trees (*cajá*, mango, *genipapo*, *samahuma*) and manioc in the case of cacao. This meant that planters could diversify their activities as they planted either of the two crops. Finally, the most delicate and time-consuming operation for both sugarcane and cacao—the harvest—had compatible schedules too: in January–February or June–July for cacao, and in March–April or August–October for sugarcane.[26]

Even more than cacao producers, sugar planters were forced to diversify their activities if they wanted the sugarcane to thrive. Intuitively, one could expect that in a region with a tropical climate of year-round high rainfall and high temperatures, insects and fungal pests would be common and would spoil the sugarcane. And indeed, in this climate any single crop planted alone in large amounts has higher chances of developing fungi and pests. When in the early twentieth century different countries experimented with large rubber plantations in the Guianas, which like Amazonia have a tropical climate, the trees quickly developed a leaf-blight disease, spoiling entire estates of planted heveas.[27] Something similar happened more recently, when the large-scale consumption and cultivation of *açaí*, the fruit of the *Euterpe oleracea* palm, led to an increase in outbreaks of Chagas disease in the Amazon region.[28] In the case of sugarcane, the planters usually cultivated cane groves far away from each other and planted numerous other crops in between. By doing so, they managed to avoid most of the sugarcane diseases that existed in the rest of Brazil, with the exception of red rot of the stem and leaf scald.[29] Whether by tradition, by wisdom, or by chance, Paraense sugar producers adopted an environmentally savvy strategy to protect the sugarcane from most of the diseases that affected their counterparts from other Brazilian provinces.

What was undoubtedly exploited in full awareness of the region's environmental characteristics was the source of energy of the largest sugar plantations: the rivers' tides. Tide mills were known since the later Middle Ages in the Middle East and Europe, and the Amazon River's tidal regime was just perfect for them. Given the almost four meters of difference between the height of the ebb and flow in the rivers near the Amazon mouth, tide mills constituted a very efficient adaptation to locally available sources of energy. When the seasonal flowing tide came twice a day, a series of either natural or human-made dams stored the water.[30] During the ebbing tide, the dammed water was funneled through an artificial water channel, spinning a water-wheel that drove the mill rollers. In the present, the archaeological remains of almost forty tide mills in the rivers close to Belém attest to the effectiveness of this technology.[31] The seasonal tidal cycle in the rivers close to the ocean, then, was not only positive for the nutrients it deposited on the floodlands but it could also be harnessed to generate mechanical energy.

The creation and operation of tide mills was not within everyone's reach, however. It required the erection and maintenance of costly hydraulic struc-tures: one or more dams, a stone or masonry channel to efficiently funnel the water propelling the waterwheel, and the machinery to convey energy to the rollers in the mill. These structures were needed in addition to the customary buildings that a sugarmill required: a shelter for the vertical stone rollers, the kettle house, a building for crystallizing and purifying the sugar, the workers' housing, and an appropriate port to ship the sugar along local watercourses.[32] Most of the tide mills, therefore, were built by the wealthiest planters and mis-sionary orders in the decades between the late eighteenth century and the Cabanagem revolt.[33] By then, the rich planters would be confronting a differ-ent set of problems.

The Cabanagem Revolt and Pará's Slaveholding Economy

Between 1835 and approximately 1840, mixed-race peasants and other nonelite groups of Amazonian society joined a series of Liberal leaders in taking over the provincial government in protest against the abuses of the Portuguese elite. The conflict originated when the arrival of new liberal ideas in Amazonia com-bined with the local peasants' unrest caused by the encroachments of slave-holding planters on their lands. To prevent the spread of subversive ideas coming from revolutionary France and to retaliate against the French occupa-tion of Portugal, the Portuguese crown planned a joint British and Luso-Brazilian occupation of French Guiana from 1809 to 1817 and established Belém as its headquarters. The larger presence of merchants, sailors, soldiers, and po-litical activists, however, only increased the dissemination of liberal and re-publicanist ideas. Intellectuals such as Franciscan monk and Portuguese exile Luis Zagalo, Portuguese-educated journalist and politician Filipe Patroni, and Canon João Gonçalves de Batista Campos organized newspapers and gave public speeches calling for the creation of a constitutional monarchy and lam-basting the persistence of Portuguese colonial tyranny after Brazilian inde-pendence. Because their arguments had a broad appeal to members of varied social classes and ethnic groups, state officials increasingly saw this political agitation as dangerous and subversive of the social order and soon started to repress it.[34] The "fire of discord" that set the whole Atlantic in flames during the revolutionary era had arrived in northern Brazil.[35]

Armed confrontations between Liberal nativists and the Portuguese faction intensified during the second half of the 1820s and the early 1830s, as did the massive mobilization and radicalization of supporters on both sides.[36] Most of the participants in these anti-Portuguese actions were poor peasants of mixed ethnic background discontented with the concentration of landowership and political power that took place during the economic expansion of 1800–1820. They were initially called *cabanos* because many lived in *cabanas* or huts near the river. Soon the protesters grew to be a heterogeneous group of Indian, mixed-race, mulatto, black, and white free but poor cultivators who embraced liberal idioms of struggle against exploitation and tyranny and adapted them to local circumstances.[37]

Going beyond the initial demands of elite Liberal leaders, the cabanos soon pursued radical social change on their own by attacking Portuguese landowners and merchants with the aim of expelling them from the region. Although not all members of the elite were Portuguese, the liberal nativist discourse soon collapsed both groups into a single enemy, coming dangerously close to equating being white with being part of the rapacious elite. Gradually, Cabanagem became a platform for members of all social groups to voice their grievances against landowners and other wealthy members of society, acquiring the colors of a social revolution.

In late 1834 political tensions escalated into war. Liberal leader Batista Campos died of septicemia caused by an injury he received accidentally while fleeing the order of arrest decreed by the pro-Portuguese governor, Lobo de Sousa. This was the last straw for the cabanos. On São Tomé Day, January 7, 1835, a force of cabano rebels took over Belém and murdered de Sousa. They denounced freemasonry and foreign domination as expressed in the political reforms coming from Rio, as well as the preeminence of Portuguese citizens in public office. The rebel forces were composed mainly of Indians, mixed-race peasants or *caboclos*, and blacks and mulattos. Their leaders were Creole merchants, planters holding middle-sized properties, artisans, and "patriot" priests. Félix Clemente Malcher, a smallholder from the Acará River region who became the first cabano president, espoused a reformist agenda, based on designating local men of property as candidates to hold public office. But he was seen as too moderate by more radical cabanos, who replaced him with Francisco Vinagre. Fearing he would lose control over the radical cabano masses, Vinagre handed the province over to the Brazilian imperial military on May 21, 1835. The newly installed imperial government quickly carried out a wave of arrests with the aim of demobilizing the cabano armed forces, which only worsened the situation. Radical cabano leader Eduardo Francisco "Angelim" Nogueira, a former tenant of the first cabano president, retook the capital city of Belém on August 23, 1835, after a week of bloody combat.[38]

Despite efforts by cabano leaders to maintain legal order and respect for private property (except for that belonging to their Portuguese rivals) throughout the interior, foreign observers and local elites soon claimed that the revolt was against "the whites." "The Indians were murdering all the Europeans," two foreigner travelers were told in Santarém in 1835. General Francisco Soares Andrea, who recaptured Belém from the rebels in May 1836 and directed the ferocious repression of the cabanos in 1837 and 1838, charged that "all the men of color born here are linked in a secret pact to put an end to all the whites."[39] German botanist Eduard Poeppig described the rebels as "hordes of robbing and blood-thirsty mestizos, mulattoes, and negroes . . . pushed on from place to place sparing only the largest towns, killing whites with indescribable cruelty and plundering and burning settlements or passing ships. Everywhere," he added, "their arrival was a signal for revolt by the coloured rabble who formed the majority of the blind workforce of the superior white Brazilian."[40] Apparently, Cabanagem had become a race war.

By 1840, the imperial government's General Andrea had defeated the rebels in every major city, but the war had wreaked havoc across the countryside. According to Andrea, in the Guajarine and Tocantins regions "most plantations and ranches were destroyed, their slaves dispersed or dead, the cattle consumed, and the most basic crops wiped out" in 1838.[41] Years later, American naturalist William Edwards narrated how "everywhere the towns were sacked, cities despoiled, cattle destroyed, and slaves carried away."[42] Other accounts of the rebellion emphasized the "immense mischief" committed by the cabano rebels, who put "many whites to death with unheard-of barbarity, and destroying their crops and cattle."[43] Estimates range between 20,000 and 30,000 dead, or about one-quarter of the Amazonian population—although more recent studies argue that this estimate may be inflated.[44] The export trade came to an almost complete halt. Not only were foreign vessels assaulted but also maritime traffic was blockaded by British, French, and Brazilian men-of-war for most of 1835 and 1836, and foreign consuls left Belém under the risk of death.

However, there are some inconsistencies in the way both Brazilian and foreign observers assessed the impact of the Cabanagem revolt on the region's economy and demography. While in September 1835 the American consul in

Pará Charles Smith discussed the palpable "fear of a massacre of the whites" during the revolt, by the end of the year he remarked that damages to American properties were "not as large as first supposed."[45] In a conversation between British naturalist Henry W. Bates and the American owner of a rice mill located near Belém in 1848, the latter explained to Bates how "when the dark-skinned revolutionists were preparing for their attack on Pará, they occupied the place, but not the slightest injury," the mill owner added, "was done to the machinery or building."[46] William Lewis Herndon, who in his travel account from 1852 emphasized the rebels' "barbarity," also explained how during the revolt "the persons and property of all foreigners, except Portuguese, were respected."[47] Except for the Portuguese community, then, the damage done by the Cabanagem revolt was probably targeted at selected opponents and more limited than previously thought.

From the standpoint of the enslaved, the revolt provided a good opportunity to escape. Numerous armed bands of plantation slaves, maroons, and deserters were formed, often breaking the barriers separating disciplined slaves from runaways. The revolted slave Francisco de Oliveira Sipião, for example, was "a captain of the cabanos and played an influential role in the disorders" of the Acará River region. He was captured and sent to Belém along with nine other slave rebels in 1836. In the same area, the "Negro Felix" maintained guerrilla operations for several years after Angelim's fall in 1835. A former slave of the Caraparu plantation, "o preto Cristóvão" led a group of at least 150 slave rebels and deserters in resisting three military expeditions sent by General Andrea in late 1835. After resisting an attack by a force of 200 soldiers, Cristóvão led the fugitives into "the almost impenetrable forest," never to be seen again. In 1838, the military commander of Muaná, in the island of Marajó, reported the existence of a group of "maroon cabanos" led by a black leader named Côco. Similar reports came from Baixo Amazonas. The ongoing war between cabanos and the Brazilian government undermined the masters' effective control of their slave crews, providing the latter with increased opportunities to escape from ranches and plantations.[48]

However, despite the black involvement in the revolt, the cabano leadership never formally abolished slavery. Far from that, cabano presidents always made it clear that their denunciation of Portuguese power was a patriotic cause in full alignment with the rule of law. Liberal leader Luis Filipe Patroni had allegedly espoused abolition in the 1820s, but no documents have been preserved showing his full and straightforward endorsement of this cause. What does exist is evidence of Patroni's project of counting the slaves in the census to determine the number of provincial representatives, along with some complaints from the slave owners in Belém that he was actually calling for abolition. Other Liberal leaders like Batista Campos had slaves themselves and did not seem to be concerned at all with abolition; they were more in tune with mainstream nineteenth-century Brazilian liberalism.[49] Therefore, despite the participation of free and enslaved Afro-Amazonians in armed actions during the revolt, no runaway slaves were to be found at the "united Brazilian encampment of Ecuipiranga," the main rebel stronghold in Baixo Amazonas.[50]

Perhaps due to the cabano leaders' failure to endorse abolition, slaveholding plantations and ranches survived the revolt. In the Acará River, the epicenter of cabano agitation, Lourenço Justiniano de Paiva's rice *fazenda* survived the armed combats of 1835–1837. Twelve out of the approximately twenty-seven slaves living there before the revolt continued to produce not only rice but also cotton and corn during those years. Another fifteen slaves from the property, nine adults and six children, were "captured" by the cabanos and ended up "in the power of the rebels"—it is unclear whether as combatants or as servants. After the cabanos were defeated, the slaves were "sent by the Governmental Legal Forces" to the State Government Palace, where their former owner eventually picked them up—a practice that seems to have been frequent in the aftermath of the revolt.[51] Paiva, then, not only kept his property functioning during the most violent years but was also able to recover some of the slaves taken by the rebels. Many other large slaveholding properties, such as the Santo Antônio da Campina sugar mill, in Vigia, the Mocajuba sugar mill in Barcarena, or the Una brickmaking factory and rice mill north of Belém, maintained large crews of slaves and went back to their regular operations after the revolt ended.[52]

Contemporary accounts suggest the possibility that some slaveowners, including the Portuguese, used their paternalistic relations with slaves to retain their loyalty during the conflict, keeping them by their side and even enlisting some to their cause. Jean-Jacques Berthier, a *babeuviste* or radical socialist from revolutionary France imprisoned in Cayenne who fled to Belém in 1820, narrated how on the eve of the rebellion the Portuguese elite used their black slaves to attack Liberal leaders. In 1824, there had been denunciations in Santarém of slaves armed by "Europeans" to fight against their Brazilian masters—although this denunciation gradually merged with other alarmed voices claiming that Indians and blacks were killing all white people. A narrative published in 1862 by French writer Émile Carrey portrayed not only a number of African slaves who remained loyal to their Portuguese owners during the revolt but also cabano rebels who kept their slaves captive and used them as canoe rowers. While we tend to associate the slaves' participation in the

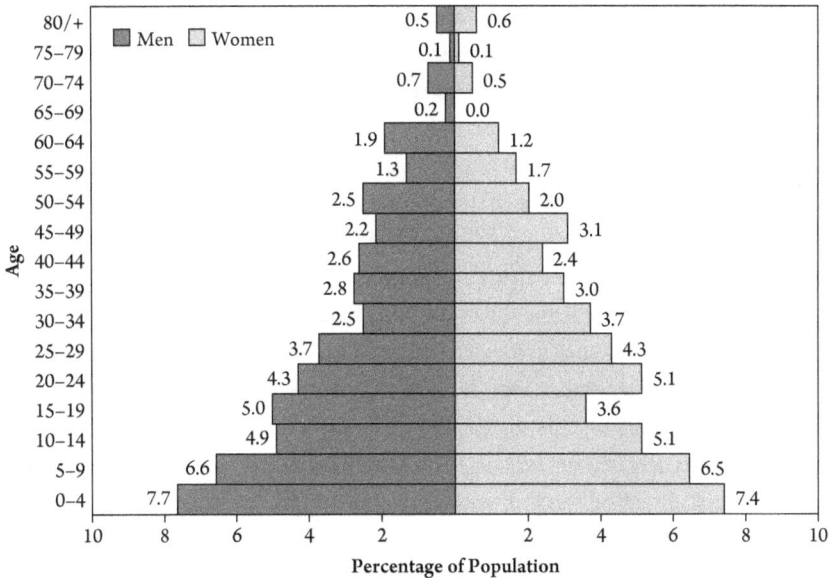

GRAPH 2 Age pyramid of rural slaves in Pará, 1850–1859. Source: Subsample of sixty postmortem inventories from 1850–1859, totaling 835 individuals with a known age. This represents approximately 2.5 percent of the average slave population for any given year in that decade. Funds APEP-Orf and APEP-JMC.

Cabanagem revolt with the liberal side, there is evidence that armed slaves did fight alongside their Portuguese masters as well.[53] At least some slaveowners, then, may have managed to keep their slave crews at bay during the rebellion.

Evidence from a collection of postmortem inventories, presented in graph 2, suggests that some adult men and women escaped or were taken from their workplaces at the time of the revolt. Normally, the age pyramid of an area with a productive slave population is thicker in the middle and narrower in the superior and inferior parts, meaning that most slaves are between 15 and 35 years old, their prime productive age.[54] The graph shown here does not follow this pattern. Among the men, the number of slaves in the 30- to 34-year-old age group—that is, enslaved males most likely either to fight or escape at the time of the revolt—was no higher than those between 50 and 54 years old and represented only about half of those between the ages of 20 and 25. The number of enslaved males older than 50 was thus disproportionately high. Children younger than 10 represented the largest population group, adding further weight to the nonproductive slave population in the province. In contrast to productive plantation economies in pre-Cabanagem Pará or in other areas, children and older adults abounded in the slave labor force.[55]

Nonetheless, the age pyramid also shows that, far from the slave population heading to extinction, by the 1850s this labor force was reproducing rapidly. Individuals younger than age 10 represent the largest age group for both sexes in the entire pyramid, reflecting a rapid increase in the number of births once peace consolidated in the region. The Cabanagem revolt, then, may have reduced the reproduction rate of the captive labor briefly, but this pause quickly gave way to an enormous expansion in the newborn and child population of enslaved individuals. The causes and implications of this expansion of family experiences among Paraense slaves are fully explored in chapter 3, but for now it seems clear that the disturbances of 1835–1840 did not suppress the natural reproduction of Paraense slaves. If anything, they just facilitated it.

Seen in perspective, this means that the revolt accelerated a demographic trend that was already present in the early nineteenth century: the creolization of the slave labor force. In the 1820s, although the slave trade to Belém was coming to a halt, at least 30 percent of all slaves in Paraense plantations were still born in Africa.[56] By the 1850s, the African-born population had dropped to only 9 percent, or 75 out of 835 individuals, according to the population sample used in graph 2. Of those seventy-five persons, twenty-four came from Angolan nations and ports (Cassange, Cabinda, Benguela, Mafundo or Mapungu, and Rebollo or Libolo); eight were from the Costa da Mina (an area comprising present-day Ghana to Nigeria); seven were from the Congo; one was Mandinga; and another one came from present-day Mozambique (Macua). Save for the 45 percent of Africans whose origins we do not know, the predominance of those from the Congo-Angola region seems clear. The Cabanagem revolt, then, accelerated the creolization of a slave population that by the 1850s had only a handful of individuals born in West Central Africa, all of them in their fifties or older.

The war also had consequences for the gender composition of the enslaved. In the late 1700s, a time when Pará received regular shipments of enslaved Africans, the population was disproportionally male on plantations and ranches: there were 149 men for every 100 women in the largest properties, 157:100 in mid-sized ones, 160:100 in small ones, and 200:100 in small ones with absentee proprietors—that is, there were two men for every single woman.[57] Between 1800 and 1835 the ratio of men declined, but remained at 120, ranging from the roughly even gender distribution in mid-sized plantations to the 141 men per 100 women in crews of ten to nineteen individuals.[58] By the 1850s, after the end of the slave traffic in the 1820s and the ensuing violence brought by the Cabanagem revolt, the masculinity rate was 98, meaning that there were slightly more female than male slaves in most properties. Cabanagem, therefore, reinforced a trend that had started by the 1800s: the balancing of gender proportions among slaves. Overall, there were 34,073 enslaved persons in Pará in 1851, representing about one-fifth of a provincial population of 179,415.[59]

In sum, Efigênia's São Miguel da Cachoeira rice plantation discussed earlier in this chapter was indeed representative of the demographic profile of Pará's slaves after the Cabanagem revolt. Having lost many individuals in prime working age, the captive population now had a larger percentage of children and elder individuals, most of whom were born on Brazilian shores, and a more balanced gender composition. During the return to normal life in the 1840s this slave population was in a good position to start reproducing itself—it had no extreme masculinity rate or pronounced ethnic barriers that would represent a significant obstacle. The enslaved had become a largely creole labor force, forming families and giving birth to children on the western side of the Atlantic Ocean, thus carving a place for their descendants in Amazonia's future.

Horizons of Destruction, Horizons of Hope

Relying on a diminished but gradually renewed slave labor force capable of producing cash and food crops simultaneously, and well adapted to environmental opportunities and constraints, slaveholding activities drove the province's economy after the Cabanagem revolt. Between 1837 and 1847 Pará exported on average almost 217 metric tons of sugar per year, for example; this amount decreased slightly to 183 from 1847–1857. In 1862, on the eve of the War of Paraguay, the 166 sugar mills of the province produced an all-time high of 540 tons of sugar in the post-Cabanagem decades.[60] Commanded by slaveowners who dominated the state legislative, sugar-producing tide mills, such as Santo Antônio da Campina, in Vigia; Mocajuba, in Barcarena; São José, in the Outeiro Island; or Jaguararí, at the confluence of the Mojú and the Acará Rivers, continued to represent symbols of wealth and power.[61]

Items other than sugar or cacao, such as rice, cotton, coffee, annatto, beef, and hides, were also exported in significant numbers between 1835 and 1870, contributing to the province's economic recovery. When outside observers talked about Pará in the central decades of the nineteenth century they did not emphasize rubber, but crops such as cacao, rice, sugar, coffee, or cotton. In 1853, American geographer Matthew Fontaine Maury called Amazonia "a rice country" and added that the region was "well adapted to the cultivation of cotton and coffee, sugar, and tobacco, of Indian corn," and other crops. While in a way he was formulating the proverbial "agricultural ideal" that observers often projected on Amazonia, he was also referring to an economic reality that

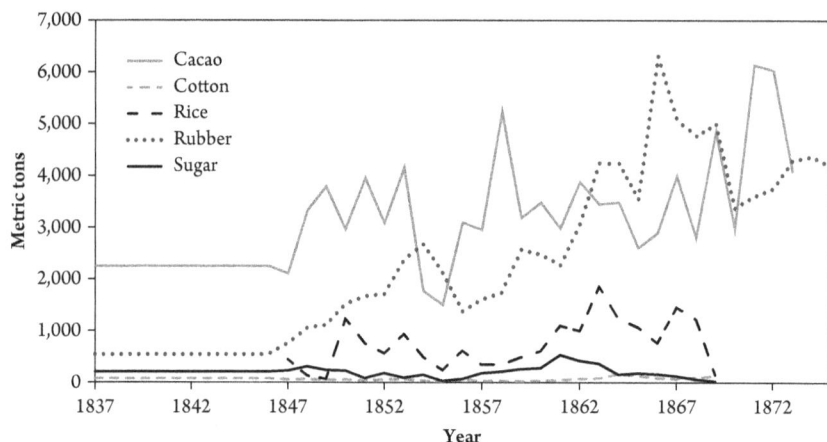

GRAPH 3 Selected exports from Pará, 1837–1875. Source: Pará, *Relatorio . . . 1858*, 34–35; Pará, *Relatorio 1867*, 16–19; Cordeiro, *O Estado do Pará*, 60.

visitors and foreign consuls often described in the mid-1800s.[62] Cacao exports, in fact, were more valuable than rubber ones until approximately 1850 and continued to be larger in size until 1862. It was only during the mid- to late 1860s that rubber took the undisputed lead in terms of size and value among all exports—until then, rice, sugar, coffee, cotton, and especially cacao continued to be central to the region's economy (see graph 3).

The case of sugar illustrates the adaptation of plantation activities to the emergence of rubber as Amazonia's first export rubric. Sugar exports started to decline in 1864 and eventually stopped by 1870. The recruitment of slaves for the War of Paraguay of 1864–1870 damaged the sugar-producing sector by liberating slaves that belonged to religious orders, especially the Carmelites, and by pushing others to flee to avoid the draft.[63] The skyrocketing prices of rubber, which doubled between 1865 and 1870, may have led capital and laboring hands away from sugar production as well.[64] But while sugar exports came to a halt, sugar continued to be produced after 1870—it was just reoriented toward internal markets, which by then were growing rapidly. President Benevides pointed out in 1875 how, thanks to the economic and demographic growth produced by the rubber bonanza, "the sugarmills do not produce enough sugar for the province," and it had to be imported from other Brazilian and international suppliers.[65] Other sugar producers switched to distilling rum, by no means an activity new to the region. Already in the late 1700s the "great consumption" of *cachaça* "in the province" and its easier and cheaper production process had led some mill owners "to prefer its fabrication over that of sugar," the provincial president explained in 1862.[66] Rum-producing mills or *engenhocas* had indeed lesser requirements than *engenhos* for capital, land, and labor.[67] Moreover, by 1875, "many" sugar plantations had become "steam-powered mills . . . with machinery to produce molasses and sugarcane brandy." Six years later Belém had 36 engenhocas, Cachoeira had 25, and Igarapé-Miri containing 116 of them—some of them employing wageworkers, but others still relying on slave labor.[68]

As this chapter has shown, that the plantation economy of Pará survived the Cabanagem revolt suggests that the destruction it caused to slave-owning estates was indeed not as devastating as most observers have argued. Many slave adults perished, were forcefully recruited, or simply escaped from their workplaces, especially the men. But by eliminating a number of adults from local plantations, the revolt also pushed the slave demography closer to that of free people, transforming the captives into a more creole, sexually balanced, and fertile population, which included a few grandfathers and grandmothers born in Africa. They were the last of their kind: only one slave ship arrived in the port of Belém after 1835, severing the direct links between Amazonia and Africa. Despite this sudden separation, Efigênia and others shared with their descendants a number of African myths and stories, for, as we see in the next chapter, those who fled their masters to form maroon communities used the image of a powerful snake as a way to narrate their saga.

CHAPTER TWO

Killing the Big Snake
Myth and History in the Trombetas River, 1800–1888

According to maroon descendant Joaquim Lima, the history of the Trombetas River maroons started "around 1800," when a number of slaves fled the
plantations and ranches of the Lower Amazon region. The runaway slaves
"navigated up the Amazon River to the confluence with the Trombetas, entering the river" and trying to find a place where they could build an autonomous social life beyond the slave owners' reach.[1] Right after entering the
Trombetas the families of Maria Peruana, old Torino, Figêna, old Preguiça, old
Ângelo, and old Carua or Caruã split from the others and headed to the Cuminá River, a left-side tributary with a number of waterfalls that could be used
as a defensive buffer.

But they were not yet on safe ground. Before arriving at the river's waterfalls, they encountered a rocky gorge about 800 meters long on the left side of
the river, which they named Rocky Shed or Barracão de Pedra. In the Barracão
there was a "big snake . . . that prevented any living being from passing through,
whether human or animal . . . even the birds flying over the place were caught
by it." The Big Snake lived in an *aningal*, a flooded area thickly grown with a
low palm known as *aningaúba* (*Montrichardia arborescens*), where she could
hide. As soon as the fugitives arrived in the aningal she attacked them. Despite
the "surprise," they fought back and "tried to defend themselves" from the ferocious animal. While "one of them was devoured by the snake . . . the others
ran away to the forest," opening a trail to the waterfalls by land. This signaled
a victory, as "above the *Pancada* waterfalls the blacks felt free from the yoke
and the persecution of the masters." Despite the continuing threat of reenslavement, they could now start a life in freedom of their own, which the slaveholding class had tried to wrest away from them.

However, the mocambeiros' interactions with the Big Snake continued. The
Snake's brother, a male serpent who lived at Lake Erepecú, repeatedly proposed
marriage to her through an intermediary—a messenger "widely known in the
region"—a dolphin named Palhão. "One day the Big Snake" felt that she had
enough, and "according to the old blacks" she "decided to respond" by visiting her brother at Lake Erepecú. A fight between them broke out, as described
by Lima:

The Snake engaged in hand-to-hand combat with her brother. Witness to their fight was an old black man who was fishing in the lake. He saw the waters moving, but could not leave the place. Hiding behind a mound, the old man waited until it was over, and then called the lake's inhabitants, who ran to its margins to see what happened. They saw the male snake from the Erepecú Lake, who survived the fight but kept only one eye, which could be seen shining in the dark at night. The elder still explain that the Cuminá's Big Snake died in combat, and one of her ribs formed the arc of Our Lady of Nazaré's altar, in Belém.

Killed in combat by her brother from the Trombetas River, who had lost an eye but had managed to survive the confrontation, the Big Snake was finally dead, signaling a pivotal event in the history of the mocambeiros: "the new families of runaway blacks could now navigate upriver from the Cuminá's mouth to the Pancada waterfall without having to cross through the forest." This made travel to and from the communities in the area easier, improving trade and communication and facilitating the arrival of new families. "With the death of the Big Snake from the Rocky Shed, the blacks" also "started to come down the waterfalls very cautiously, building new homes." The Rocky Shed, the cavity the Big Snake used as a shelter, now became a place with great symbolic importance for the mocambeiros. Thankful for being able to move around freely, they "made promises to the Holy Heart of Jesus, whose image was carved out by nature in the upper part of the Shed's rocks." Punctuated with images of place, nature, and religion, the story of the Big Snake's death signaled the mocambeiros' relocation to new residential spaces below the waterfalls, where they would fully enjoy their hard-won freedom.

Joaquim Lima narrated this story in 1992, at a time when the communities of the Cuminá River were threatened by the construction of a hydroelectric plant.[2] With the help of Liberation Theology priests, NGOs, black political organizations, and some anthropologists, representatives of the maroon villages from the Cuminá and Trombetas Rivers met in the city of Oriximiná to form the Quilombo-Descendant Association of the Municipality of Oriximiná (ARQMO), the first step in the application for a collective land deed from the federal government. This was the first time that maroon descendants pushed the Brazilian government to actualize the recognition promised in the 1988 constitution. A great-grandchild of Maria Peruana, one of the first maroons to make it to the waterfalls, Lima self-identified himself as a "maroon descendant" (*remanescentes dos quilombos*) fighting "for the demarcation and the grant of land deeds" to "guarantee the freedom we have conquered."[3]

Clearly, Lima's goal in telling this story was to underpin the mocambeiros' claims to land and recognition. However, the fact that this oral tradition contains "a particular purpose and fulfills a particular function," in this case legitimizing the claims to ancestral occupation, does not render Lima's version of the Big Snake useless. Accustomed to dealing with oral traditions in lieu of historical documents, Africanist historians have suggested some methods to use them as historical sources. First, most symbols in this narrative are widely shared among the maroon descendants of the Trombetas.[4] While some details in the story vary among its tellers, the core of the Big Snake narrative remains consistent. Second, I tried to place the narrative in conversation with "outside sources that can be checked and certified as independent," such as police and governmental reports, travel accounts, genealogical trees, and interviews with the mocambeiros. With this I seek to generate a dialogue between narrative and written sources and to dig as deep as possible into its key natural and topological symbols.[5] I considered the tradition of the Big Snake a "hypothesis," a source that could enter into dialogue with, and even correct, "other perspectives just as much as other perspectives [could] correct" it.[6]

"One of Them Was Devoured by the Snake . . . the Others Ran Away to the Forest"

The Trombetas River originates in the Guiana highlands, near the border with Guiana and Suriname. On its way to the south it receives long tributaries like the Mapuera, the Cachorro, and the Cuminá, and eventually joins the Amazon River about halfway between the cities of Manaus and Belém. Its course is punctuated by a number of waterfalls or *cachoeiras*, which made navigation very difficult during the colonial period and discouraged European exploration and settlement. Indian nations populated the Trombetas River area before the arrival of the Europeans: the Carib-speaking Kaxúyana, Pianocotó, Parikotó, and Tiriyó Indians lived in its headwaters, and groups of Tupian-speaking Mura, Maués, and Mundurucú inhabited its lower course.[7] Overall, the middle and upper course of the river remained poorly known until the nineteenth century.

During the colonial period the Portuguese Crown's agenda for the region of Baixo Amazonas consisted of populating it effectively while fostering agriculture and the collection of forest spices. During the seventeenth century a series of Capuchin, Franciscan, and Jesuit missionaries founded the settlements that would eventually become the present-day cities of Santarém, Alenquer,

Trombetas

Poana
(1855)

Erepecurú

Turuna
(1855)

Maravilha
(1855)

Quilombo to Torino

Curuá

Santa Luzia

Figênia
Santa Anna
(c. 1877)

Conceição
(1871)

Acapú

Stone Shed

Jacaré
(1909)

Lake
Erepecú

Tapagem
(1875, 1878,
1898, 1909)

Mamiá

Mãe Cué
(1914)

Moura
(1911)

Inferno
(1854)

Jamarí   Juquirí
(1917)   (1875)

Oriximiná

Pacoval

Alenquer

Óbidos

Terra Santa

Faro

Juruti

Amazon

Santarém

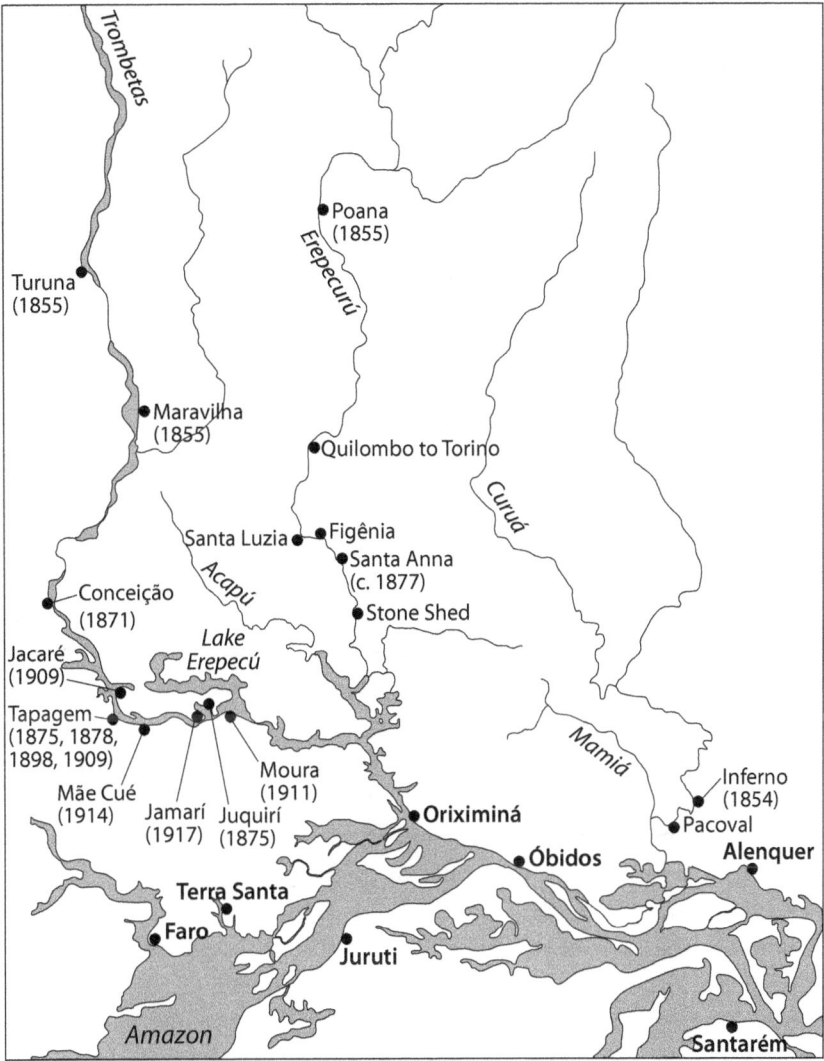

MAP 2 Maroon communities in the Lower Amazon, pre- and post-1888. Sources: Pre-1888 locations are based on Andrade, "Os Quilombos da Bacia do Rio Trombetas," 88–91; Azevedo, *Puxirum*, 92; Coudreau, *Voyage au Cuminá*, 21–22, 55; Cruls, *A Amazônia que eu vi*, 61; Derby, "O Rio Trombetas," 370; Funes, "Mocambos do Trombetas," 244, 246; Funes, "Nasci nas Matas," 172; Lima, "História dos Negros," 4–8; Ruiz-Peinado, "Maravilla," 118; Conselho Nacional de Proteção aos Índios e à Agricultura, *Diário das Três Viagens*, 18, 35. Post-1888 locations are from Brown and Lidstone, *Fifteen Thousand Miles*, 276; Coudreau and Coudreau, *Voyage Au Trombetas*, 16; Derby, "O Rio Trombetas," 369–70; Ducke, "Explorações Scientíficas," 159–60; Funes, "Mocambos do Trombetas," 237, 244, 247, 251; Rodrigues, *O Rio Trombetas*, 20, 121.

Monte Alegre, and Óbidos, which was the closest one to the Trombetas River. By the late seventeenth century the Portuguese Crown grew concerned by the rising influence of other imperial powers, and to better guard the river it erected an artillery battery in Óbidos (called at the time Pauxís), a strategic location given the Amazon River's narrow width passing the city.[8] In 1758, the new directorate system of civilian supervision of the Indians brought further changes: civilian administrators replaced the missionary orders, and the mission of Pauxís became the town of Óbidos. The settlement grew significantly in the following decades, and in 1798 Bishop Frei Caetano Brandão described Óbidos as "one of the most opulent populations of the state," with about 900 souls including Indians and whites.[9]

By then the spread of cultivated cacao was altering again the Lower Amazon's demographic and economic profile. In the mid-1700s the population was primarily composed of indigenes and a fast-growing class of *mestiço* peasants; the arrival of enslaved Africans in large numbers in the mid- to late eighteenth century added a new culture to the local mixture. In 1820 the area had become a prime cacao-producing center, propelling Pará's exports of this item to unprecedented levels.[10] During an 1819 visit, British naval officer Henry Lister Maw observed how "the cocoa plantations now extended for miles along the banks of the river," including a very large one owned by the emperor.[11] By then 1,294 of the city's 4,281 inhabitants, or 30.2 percent, were enslaved black persons.[12] Other cities of the Lower Amazon region came to have equally high percentages of slaves, most of whom were working on cacao plantations; others cultivated tobacco, sugar, and foodstuffs, or worked on the few livestock farms and sawmills that existed near Santarém.[13]

But while the area was known for its cacao production, the many lakes and canals of the Lower Amazon, as well as the rugged topography of the left-side tributaries of the Amazon like the Nhamundá, the Curuá, or the Trombetas, made Indian and slave flight a chronic problem.[14] As early as 1799 the authorities reported that a "formidable *mocambo*" of Indians and slaves was causing trouble for cattle ranchers in the Trombetas River area. Joaquim Lima's narrative puts the arrival of runaways to the area somewhat later, "around 1821."[15] According to both contemporary and present-day studies of the region, the movement of Mura indigenous groups away from the mid- and lower course of the Trombetas River did take place at the same time as the settlement of "maroon groups" in the area.[16]

A wave of armed confrontations between runaway slaves and local planters took place between 1801 and 1827, when the municipalities of Óbidos, Alenquer, and Santarém financed a series of attempts to recapture the runaways and quash their settlements. After sending two groups of armed men in 1801 and 1805, in late 1812 the three municipalities joined forces again to form a military force of almost 400 men guided by Mundurucú Indian rangers. This large contingent headed to the mocambos of Cipotema, Caxange, and Inferno, capturing fifty-nine adults and fifteen children—although a number of them eventually escaped to the forest. In 1823, a new military expedition disbanded a mocambo in the Trombetas and captured Atanásio, a *carafuz* (person of mixed Indian and African descent) runaway slave who appears to have lived among the Curuá River maroons ten years before. In 1827, the authorities sent yet another expedition to the area, "capturing many slaves," but "some always escaped, continuing to resist in the forest."[17]

These anti-maroon activities shed light on the Big Snake story. As recounted by Joaquim Lima, the Big Snake employed physical violence to prevent the runaway slaves from reaching a "safe haven" above the waterfalls. "Even the birds flying over" the Rocky Shed "were caught by it" and would not go beyond the home of the threatening ophidian. She represented a fierce, violent, and potentially mortal force opposed to the mocambeiros' attempt to create a space of autonomy. The beast indeed managed to kill one of the mocambeiros, a terrible metaphor of the risks they ran in their escape to the waterfalls. In this sense, the Big Snake seems to symbolize the military might of the slaveholding elites, who tracked and killed some mocambeiros in 1801, 1805, 1812, 1823, and 1827—just like the mythological ophidian. Merging the challenges of exploring new environments with the fear of the power of the slaveholding class, the metaphor of a Big Snake expressed the dangers faced by the mocambeiros as they opened up new spaces of autonomy in a world controlled by slaveowners.

It is not surprising that the mocambeiros chose a snake to symbolize the threatening power of the slaveholders. Snakes appear often in West and West Central African folklore—and consequently in the folkloric traditions of Afro-descendants across the Americas. Snakes were also an important divinity among Haitian vodun practitioners from Benin, and among the practitioners of Yoruba religion in Cuba and Brazil. They appear in numerous oral traditions from Georgia, Louisiana, Venezuela, and many other places.[18]

Among Bantu-speaking peoples from Congo and Angola, who constitute the majority of the Africans who came to Amazonia, snakes appear in a variety of narratives where they take on a number of different roles, from kidnappers of women to carriers of the spirits of the death.[19] In the early twentieth century, for example, residents in the northern region of Congo believed that a powerful spirit known as Mpulu Bunzi inhabited the rivers and had power over floods and rains. In a fratricidal confrontation, Mpulu Bunzi, a female, confronted her male brother Mpangu Lusinzi and ended up killing him. While the genders of the two protagonists are inverted, the story is analogous to that of the Big Snake. She was also associated with "extraordinary rocks," a strong similarity with the mythical qualities of the Rocky Shed. When the simbi, water spirits that guarded the natural world, came from Congo to South Carolina along with other African spiritual traditions, they continued to adopt numerous animal and mythical forms, including mermaids and snakes.[20]

In the Angolan region of Mayombe there is another oral narrative about an enormous snake called Mbumba Luangu, who represents the rainbow. It "comes out of water and wriggles up the nearest high tree when it wants to stop the rain," and is a very dangerous entity: "anyone who sees it—that is, sees the place where its end seems to rest on the earth—runs away as fast as he can: "if he sees you he will kill you."[21] The common elements with the Big Snake of Lima's narrative are clear: the aquatic life, the power to affect other natural elements, and the dangerous nature of the animal appear prominently in both narratives. It is very likely that the maroons resorted to symbols they had known in Africa to try to make sense of life in Amazonia.

But the Big Snake narrative was also influenced by the oral traditions of the Kaxúyana Indians from the Trombetas River, who over time established a full spectrum of relations, from collaboration to confrontation, with the maroon newcomers. The Kaxúyana, a Carib-speaking group from a Trombetas tributary known as Cachorro, have a myth of origin based on the story of Purá and Murá, the two original humans who lived at the headwaters of the Cachorro River and fabricated all other humans. In the beginning Purá and Murá made some persons out of a wood known as *pau d'arco* (*Tabebuia impetiginosa*), gave them a canoe, and ordered them to navigate downriver to populate new areas. Purá also warned them that a dangerous snake called Marmarú-imó lived in the waterfalls of the Cachorro. The humans made by Purá and Murá left on their way to a new home, but as they approached the waterfalls the treacherous snake opened her giant mouth and gobbled them. To avoid losing more humans to Marmarú-imó, Purá and Murá decided to kill her. Armed with knives, they went to the waterfalls, let the snake gulp them and, once they were inside her stomach, "they slit the snake from head to tail": Marmarú-imó was now dead. After inspecting the dead body, the two heroes learned a number of designs that were on the snake's skin and her ribs that they later used to decorate domestic utensils, such as baskets and sieves. The painted designs from one of her ribs, for example, were used to decorate the *tipitis*, the utensils used to squeeze the water out of boiled manioc.[22] As can be seen, this narrative probably had a strong influence on the maroons' legend and probably constitutes its origin.[23]

In the early 1830s, the political instability that would soon lead to the Cabanagem revolt spread across the region, engulfing the confrontation between masters and runaway slaves in the larger one between the rebels and the governmental forces. Military operations in the region were substantial, and the number of runaways seems to have increased in this period. However, the cabano leaders did not admit runaway slaves into Ecuipiranga because, as I described in chapter 1, they claimed to be defending the essence of the Brazilian nation and the legal pillars of the Brazilian state—slavery among them. In the Lower Amazon the revolt-turned-war ended only after the rebel strongholds of Ecuipiranga and the Curuá River were defeated in 1838. But the respite that the Cabanagem revolt paradoxically brought to the maroons could not last for long, because the Big Snake was still alive.

"Above the Pancada Waterfalls the Blacks Felt Free from the Yoke and the Persecution of the Masters": Maravilha and Other Maroon Settlements in 1855

As the economy recovered and revenue from exports slowly flowed back to the state's coffers during the 1840s, Paraense elites launched a new wave of coercive measures against the free poor population in an effort to reassert the power of the state. The implementation of this agenda started with the creation of the Corpo de Trabalhadores (Labor Corps) in 1838, which had the goal of "removing from idleness an excessive number of Indians, blacks, and mestiços lacking education, who exceed three quarters of the population." Poor nonwhites who could not prove they had a stable job or landed property were forced to enlist in the Labor Corps, where they were put to work for private and public companies.[24] Enrolling some five thousand workers the Corps functioned until 1859, when it was abolished because of its tendency to be "a mechanism of gain" for the officials in charge of managing it. In its crusade against the autonomy of popular groups, the state government also targeted the itinerant retail merchants or *regatões*, whose activities allowed rural populations to circumvent the monopoly of landed elites over wealth and the supply of manufactured goods—and sometimes to avoid taxation as well. The regatões' activities were banned in 1850, although so many protests arose from the interior that the prohibition was struck down in 1854.[25]

In this oppressive milieu, the government of Pará also declared war on runaway slaves. Seeking to quell the threatening "evil" of marronage,[26] in 1854

President Sebastião do Rego Barros requested that the municipal chambers reported on the number and size of the mocambos present in their territories, and provided new funds for hiring *capitães do mato* or maroon-hunting specialists.[27] "We know that in this district there have been quilombos in the upper river for many years," Óbidos police chief Auzier explained to Commissioner José Joaquim Pimenta de Magalhães. "Since 1840 ... more than 150 [individuals] of both sexes" had ran away from Santarém and Óbidos "to join those who already existed in the quilombos, whose number we cannot ascertain because they date from long ago."[28] This intolerable scourge demanded swift action.

The following year, state and local authorities assembled a military expedition against Maravilha, a large maroon settlement located in an island above the waterfalls. The armed force of about 190 men, including regular troops and Mundurucú rangers led by their *tuxaua* or chief, left Óbidos for the Trombetas in October 1855 under the command of militia captain Maximiano de Souza.[29] The participation of the Indian rangers was seen as fundamental to the success of the endeavor. As the Santarém city council explained in 1862, the rangers would detect "any vestige [of human activity] in a small bush leaf, [finding out] who had passed by the place[,] or whether any human being had actually done so, and whether somebody lived nearby."[30] But despite the presence of the Mundurucú rangers in the expedition, success was far from guaranteed. For four days the governmental forces played cat and mouse with individual fugitives before reaching the waterfalls; in the process forty-eight soldiers deserted because of the hardships of fighting in mocambeiro territory. On the fifth day they neared the waterfalls and captured the elderly maroon Benedicto, who told them the precise location of Maravilha and some information about the seventy individuals of both sexes living there. Commander de Souza led the troops to the encampment, only to find it deserted. Benedicto's travel companions, who had not been captured, had run to the settlement to warn the mocambeiros of the expedition, and the runaways had all escaped to the forest. Benedicto himself would escape his captors shortly afterward during a stormy night. Once again, the maroons seemed to slip through the grasp of a slave-catching expedition.

The mocambeiros' successful resistance was based on the acquisition and effective use of knowledge about the natural world of the Trombetas River. The Big Snake myth characterized the waterfalls as the gateway to a world relatively safe from the slaving forces—and this seems to have been the case. In his report of the attack of Maravilha, Maximiano de Souza had explained that "to clear the waterfalls you need a canoe adapted to this end, manned by a special

pilot and with its baggage prepared to this exceptional way of traveling"—
something hard to come by. In April 1854, the Alenquer police chief had explained to the Pará commissioner how "you cannot arrive" at the Mamiá
quilombo "by canoe due to the waterfalls," which made it easy for the mocambeiros to defend the access routes.[31] Even after clearing the cachoeiras, the
enslavers faced sophisticated systems of vigilance involving skilled mocambeiro rangers who would detect the approach of enslaving troops, a number
of "spies" who would make any expedition "fruitless," and special drumming
codes used to warn other communities that a hostile force was approaching.[32]

Not even the way back home from mocambeiro-controlled territory was
safe for the troops. In 1852, "almost all the soldiers in the expedition plus some
of the oarsmen" sent from Santarém to recapture a group of slaves who fled to
the Trombetas fell sick suddenly, as did a number of Maximiano de Souza's
troops on their way back home when they caught "a bad fever."[33] That illness
was actually not a fever, but resulted from intoxication caused by drinking water
poisoned with *timbó*, a vine used by the Kaxúyana and other Indians in fishing. In the words of American entomologist Herbert Smith, "the fugitives know
the country, but the soldiers are utterly ignorant of it," so "each expedition" to
the Trombetas "has returned with fever and disappointment." The maroons
used waterfalls, *igarapés* (streams), vines, and multiple other natural elements
as weapons. They had, in other words, weaponized nature.[34]

The narratives of Maravilha and other expeditions indicate that these
maroon communities were large, although it is difficult to know their precise
size. In January 1854, the Óbidos police chief calculated that "more than one
hundred and fifty [maroons] of both sexes" lived in the Trombetas area; three
years later a Portuguese explorer claimed that they were "no less than 300."[35]
In his report, de Souza explained that in Maravilha there were thirty-six
palm-thatched huts built with daub on latticed frames of wood, and he calculated a total of seventy inhabitants for the settlement—which seems low. A
more reliable estimate, based on the high fertility rate of the rural population
and the 172 runaways living in the Inferno mocambo on the Curuá River,
would be that between 120 and 200 individuals lived in Maravilha.[36] If we add
other maroon groups on the Trombetas and the Cuminá, the total on both
rivers could range between 400 and 800 maroons. Such an estimate is close to
the Santarém council's calculation of "three to four hundred of both sexes and
different ages" in 1862 and to Orville Derby's calculation of "several hundred"
in 1871, but far from Tavares Bastos's exaggerated figure of 2,000 in 1865.[37] In any
case, these are high numbers for a municipality that had 1,138 slaves and 8,970
free inhabitants in 1872.[38]

FIGURE 2 Figêna, a spiritual leader of the Cuminá River mocambos, in 1900.
Coudreau, *Voyage au Cuminá*, 32. Courtesy of University of Illinois at
Urbana-Champaign Library.

Considering the history of permanent aggression against them, the most
surprising feature of the Trombetas maroon society is the presence of numerous families among them. Joaquim Lima noted that the earliest slaves who fled
to the Trombetas were from the families of Maria Peruana, old Torino, Figêna,
and others.[39] It is possible that Lima was talking about families just as a way
of naming certain individuals—which is still common practice in the area. Yet,
the reports of the enslaving expeditions in those years do confirm the presence of numerous families in the mocambos. The report of the expedition
against the Cipoteua, Inferno, and Caxanje mocambos in 1813, for example,
states that, of the eighty-three residents (thirty-three women and fifty men)
in those settlements, as many as nineteen were children. In 1854, a police report from Óbidos noted 150 maroons "of both sexes" living in the Trombetas,

and a travel account produced shortly thereafter mentioned seventy-five maroons (forty-four men and thirty-one women) living in the municipality of Óbidos. Of the 135 Curuá mocambeiros who were sent to Belém in 1876 after turning themselves in, forty-four individuals, or about one-third, were younger than ten years old.[40]

Given the challenge of hiding from the slave catchers, the risk of armed confrontations, and the possibility of recapture, marronage has traditionally been seen as a fundamentally masculine endeavor, with a few exceptions in the Caribbean.[41] More recent analyses have modified this picture by emphasizing how "the ties binding mother and child were often not a deterrent but rather a powerful motivating factor in women's decisions to flee, regardless of a child's age."[42] Thus, a maroon community would be seen as a desirable place to have children and form a family, in spite of all the potential dangers. The story of Margarida, a mulatto slave from Santarém who fled her owners in 1851 to seek refuge in the Curuá River mocambo, is a case in point. Between her arrival at the mocambo in 1852 and the year of 1868, Margarida and her maroon partner Bernardo Antonio Ferreira de Christo had six children: Maria Olimpia in 1855, Maria Francisca in 1857, Maria Bernarda in 1858, Maria Luiza in 1859, Manoel Pedro in 1865, and Manoel Lino in 1867.[43] We do not know whether Margarida ran to the Curuá with the express intent of forming a large family, but at the very least she found there a supportive atmosphere to do so. Joaquim Lima's Big Snake narrative, therefore, seems accurate in its claim that family groups were central to the Trombetas River maroons. If they could slip past the Big Snake, then finding the emotional solace of a nuclear family was a realistic expectation.

There is one last but important implication of the presence of families in the mocambos. Such a large, growing population could not have subsisted just by "committing robberies" of local farmers and merchants, as provincial presidents claimed time and again.[44] Instead, these nuclear and extended family units cultivated the omnipresent manioc, along with beans, rice, and vegetables, and collected fruits from the planted and wild fruit trees of Amazonia.[45] Thus, in 1831, an anti-maroon expedition found cultivated manioc plots and "three ovens to make manioc flour" in a quilombo near Santarém, and in 1848 the governor of Pará explained how a mocambo-hunting expedition had found "an establishment with 59 abandoned huts," as well as "manioc grounds, large sugarcane fields, and other planted items nearby," including "utensils to make manioc flour and sugarcane rum." In 1877, missionary José Nicolino de Sousa visited the maroon villages of the Cuminá River, noticing the "roças" and the "bananas and the sugarcane" that many mocambeiros cultivated in the forest.[46]

Given the size and the family composition of maroons in the Trombetas and other Lower Amazon rivers, agriculture was not an option, but a necessity.

The presence of numerous families also meant that the communities needed to develop ties with the outside world. In May 1856, after the failed attempt to destroy Maravilha, Governor do Rego Barros bitterly wished his successor "that you may be more successful than I have been . . . having more resources and finding less support and protection [for the maroons] from some neighbors that communicate with them, whether from fear or from sordid self-interest." This communication enabled the maroons to be "soon warned against" the authorities' attempts to recapture them, "thus beating the expeditions most of the time," lamented another governor in 1845.[47] The maroons had external allies, and they came to play a prominent role in creating landscapes of peasantization below the waterfalls. But first, a ferocious ophidian needed to be defeated.

"With the Death of the Big Snake . . . the Blacks Started to Come Down the Waterfalls": Abolition and the Move below the Waterfalls

The way the Big Snake story goes, after the dolphin Palhão took her brother's marriage proposal several times to the Big Snake, she "came out of the Rocky Shed" to confront her brother at the Erepecú Lake. Once there, they "engaged in hand to hand combat," resulting in the demise of the snake. "With the death of the Big Snake from the Rocky Shed, the blacks started to come down the waterfalls very cautiously, building new homes" closer to the Lower Amazon cities. This narrative displays joy and optimism: as mentioned, thankful for His favors, the mocambeiros "made promises to the Holy Heart of Jesus" that nature had carved in the upper stones of the shed, and entered a new stage in their history.[48]

At first glance, the Big Snake's death seems a fitting metaphor for the abolition of slavery, which took place in Brazil on May 13, 1888. However, while in the Trombetas the locals indeed celebrate the promulgation of the Golden Law on May 13, the demise of slavery is not seen as a one-day event. Rather, it is usually recounted as a gradual movement from the safe havens above the waterfalls to new areas below them, a decades-long migration that led to free mobility and an easier life closer to the cities of Óbidos, Alenquer, Santarém, and Monte Alegre. Accordingly, the beast's demise stands in for the gradual erosion of the slaveholders' power, not for the passing of the Golden Law in 1888. As oral historians have shown, key points of a narrative such as the Big Snake's death can become a "place of memory," a powerful symbol condensing entire sets of personal, collective, and institutional changes.[49]

The mocambeiros' relocation process was catalyzed by a series of political and economic changes that took place in Pará in the 1860s and 1870s. During those decades rubber took the lead among the imperial province's exports in terms of value and volume, ushering in a new era when extractive products dominated the Amazonian economy and new commercial elites came to power—sometimes in alliance with old ones.[50] Gradually, both poor peasants and landowners from the Lower Amazon cities started settling near the little-known Trombetas and other rivers, looking for natural riches (turtle eggs, Brazil nuts, vegetal oils like *copaiba* or *andiroba*), as well as pastures and suitable agricultural lands. The 1850s wave of maroon repression discussed earlier reduced the frequency of confrontations between maroons and the Óbidos slaveholders, and the Trombetas River area started to be seen as relatively pacified. Pará's government, in turn, funded a number of studies and expeditions exploring new venues for economic growth in areas heretofore poorly known.[51]

In these circumstances, President Joaquim Raimundo de Lamare took office in 1866 and considered trying a different approach to solve the problem of marronage: sending missionaries. Thus, in 1867, Capuchin Father Carmelo de Mazzarino, convinced that "the only mighty force that could reach the mocambeiros is the Gospel," was sent by the Pará government to the Trombetas to establish a permanent mission to pacify them.[52] Missionaries had been used before as intermediaries between authorities and maroons in other corners of the Atlantic world: the establishment of missional systems had been for centuries the most frequently used strategy to deal with unconquered Indians, so colonial authorities found this strategy a natural and appropriate one to deal with maroon groups everywhere they existed, from Florida to Colombia.[53]

Mazzarino had a brief "probationary" stay with Florenciano, a black farmer living below the waterfalls, who was in continuing communication with the settlements upriver. After arriving at the waterfalls of the Trombetas, Mazzarino was allowed to enter those settlements, despite the mocambeiros' initial reluctance. There he observed the maroons' complex systems of security, including the secret drumming codes that signaled the arrival of travelers at the Porteira waterfall, the entrance into maroon-controlled territory. While Mazzarino was unable to establish a permanent mission, he did baptize numerous children and married many adults at the Campiche igarapé, near the ruins of Maravilha, where the mocambeiros had constructed a small chapel. He advised them to move downriver below the waterfalls, but only a few did so because of the still widespread fear of reenslavement.[54]

Mazzarino also served as an intermediary between the maroons and the provincial government in the negotiations over the legalization of their status. Aware that during the War of Paraguay of 1864–1870 emancipation had been promised to those who joined the Brazilian army, the maroons agreed to become regular citizens under two conditions.[55] First, they would be exempt from recruitment for the war. Second, they would pay 300 *milréis* over four years in exchange for their manumission—with the elderly paying a lower amount. Were these conditions not respected, they threatened to resettle in Suriname, where slavery had already been abolished. However, the provincial government's policies were far from consistent throughout the empire, with provincial presidents changing even within a single year. Thus, the tides turned on October 31, 1870, when Conservative Party leader Canon Manuel José de Siqueira Mendes signed Law 653 authorizing the destruction of all mocambos in Pará, probably due to the pressure exerted by the slave-owning elite.[56]

Just six years later, Father Nicolino de Sousa conducted three expeditions to the Cuminá River to locate the large prairies that supposedly existed above the waterfalls, the so-called *campos gerais*. His mission implied "the catechizing and reduction of the Indians of the region," and to achieve these goals, he needed to rely on the most reliable sources of information and guides to navigate the river, who were, no doubt, the mocambeiros.[57] Like Mazzarino, Nicolino was able to freely navigate the Cuminá, and he eventually visited the maroons, celebrating mass and employing some of them as pilots to navigate the waterfalls. He met Old Toró's daughter Maria de Jesus and slept at the house of Old Lautério, two maroons who were among the founders of the communities, according to Joaquim Lima's Big Snake Narrative.[58] Eventually, Nicolino succumbed to malaria after returning home from a third expedition in October 1882.

Both Mazzarino and Nicolino confirmed that the maroons were the most knowledgeable guides to the Trombetas and Cuminá Rivers. They were skilled canoemen who could clear the waterfalls by sailing across them, as French explorer Otille Coudreau witnessed some years later. In her 1900 exploration of the Cuminá she hired Old Santa Anna, the "famous" and "indispensable" guide who drove Father Nicolino and other explorers to the lands above the waterfalls during the 1880s and 1890s.[59] The maroons could also bypass the waterfalls by finding *rochedos* or paths through the thick forest. Moreover, they were invaluable participants in these expeditions because they knew where to buy manioc flour from relatives and friends, where to find groves of edible plants (açaí, Brazil nuts, oranges, bananas, mangos), where to hunt, and where to locate a *tapera* or abandoned agricultural location, a good place for setting up camp.[60] After decades withstanding the slaveholders' attempts to disband their communities and capture them, the mocambeiros used their advanced knowledge of transportation and survival in the wild to become arbiters of the area's exploration.

The expansion of clandestine commercial relations between regatões and the mocambeiros would make this role even more evident. In contrast to the case of the missionaries, however, the merchants' activities were beyond the authorities' control. The Óbidos police chief explained to state authorities in 1854 that "during May and June the maroons descend" from the mocambos "to purchase gunpowder and weapons." "Taking advantage of the people's coming to the city during the holidays," he reported again in January 1857, "the Trombetas' blacks . . . had come for supplies."[61] And if the maroons did not go downriver, then the itinerant merchants would visit them: "there is not the slightest doubt that they do business with slaves[,] especially with those living in quilombos along the Trombetas River," Councilman Ambrosio de Andrade Freire denounced the following year. An 1862 petition from a Santarém councilman for anti-maroon expedition funds explained how, thanks to the merchants, the "blacks" had access to "all they need, including cloth, salt, gunpowder, and weapons." Even Brazilian naturalist Ferreira Penna was able to observe how in Óbidos "the largest amount and the best quality" of tobacco "comes from the Trombetas River maroons." His source for this information was the "regatões and other pilots who, in the pursuit of commerce, have reached the waterfalls and even beyond."[62]

By the 1870s this illicit trade was conducted, according to numerous observers, "in broad daylight." Botanist João Barbosa Rodrigues explained how the maroons "started to come, even during the day and in sight of the authorities, and they not only buy and sell but also bring their children to the parish to be baptized, boldly stating that they are mocambistas." Rodrigues saw "some of their canoes" at the Óbidos port, adding that "some [maroons] stayed in my house during the day. It is no longer surprising to see them disembarking in full daylight. What *is* surprising," he added, "is to see how they meet with their [former] masters, asking for their blessing and leaving in peace."[63]

Taken together, the increasing contacts with missionaries and merchants speak of a spatial confluence between the mocambeiros and the broader Amazonian society. On the one hand, the demise of slavery and the gradual expansion of the Paraense economy in these decades stimulated the exploration of poorly known areas—and the mocambeiros' mastery of the wilderness came in very handy in the Lower Amazon rivers. Forest guides able to navigate the waterfalls and to locate Brazil nut groves proved very useful to the commercial houses sprouting all over the Amazon basin thanks to the rubber bonanza. On the other hand, the maroons' proposal to settle as free men and end hostilities with the slaveholders indicates that, after years of armed resistance, there was a willingness to achieve peace as long as they could maintain their autonomy—a logical desire considering the past hardships of resisting military expeditions while raising entire families. Witness to this desire is the number of baptisms that Fathers Mazzarino and Nicolino performed at the mocambos. While the Big Snake died fighting her brother, the threat of reenslavement passed away gradually, as the mocambeiros established further ties with the regional society and relocated closer to populated areas during the 1870s.

Abolition, Migration, and the Death of the Big Snake

The narrative of the Big Snake is an account full of rich symbolism of the maroons' struggle against the slaveholders' power. In the first part of the story the beast constituted an obstacle in the struggle to erect peasant communities. The Snake's Rocky Shed represented a locus of power, because it guarded the path to the natural defensive buffer of the waterfalls. This protection was indeed fundamental for the maroons, as not one but two waves of repression threatened their very existence: one occurring before the Cabanagem revolt and another one in the 1850s. Once deciphered, the narrative reflects perfectly what happened when the runaways reached the Rocky Shed: either they were hunted down by the Snake (the pro-slavery forces), or they managed to pass through it, reaching the safe havens that existed above the cachoeiras.

In the second episode of the Big Snake narrative, she came out of the Rocky Shed to confront her brother and eventually died fighting him. The beast's death took place at a time when Pará's extractive economy was booming, leading the state government to provide support for the expansion of economic activities along its multiple unexplored frontiers. One of the Snake's ribs went to the altar of Our Lady of Nazaré, the patron saint of Amazonia, and the Holy Heart of Jesus appeared naturally at the Rocky Shed on the ophidian's death. These narrative elements express religious habits and beliefs that newer state elites exploited by sending missionaries to the communities to pacify the area and tap into its riches. Simultaneously, the maroons themselves started to take advantage of the presence of ruthless itinerant merchants to establish commercial relations that, considering the absence of competing Brazil nut gatherers in the Trombetas at that time, were probably lucrative—or at least provided stable access to cash. The maroons experienced the ongoing process of abolition at the national level as a migration down the waterfalls, remembered through the episode of the ophidian's death.

The mocambeiros used a series of natural metaphors to express the obstacles and the victories they achieved from the moment they fled their masters to form free communities. Using fragments of West African and Amazonian mythology, they assembled a rich historical narrative expressing both the estrangement created by forming social bonds in a new place and the use of local environments in designing mechanisms and strategies to protect their families and their household economies. The Snake, the waterfalls, the Rocky Shed, and the Holy Heart of Jesus carved on its front show how the maroons embedded their story in a specific landscape: the different sites appearing in the narrative constitute pivotal fragments in this story of violence, endurance, and triumph. As Jan Vansina argues, "the views one has about space influence one's view of past happenings . . . the importance of orientations like these can scarcely be exaggerated."[64] The story of the Big Snake uses the river's natural geography to sustain an interpretation of the maroons' origins that emphasized autonomy and community.

The Big Snake narrative also bears witness to the sheer centrality of the natural landscape for the viability of marronage in Amazonia. The success of the mocambeiros' efforts to resist the Snake was determined by their capacity to use the waterfalls as a proper defensive buffer, to find escape routes in the forest, to create networks of informants, and to weaponize nature with the intent of slowing down or even defeating the slaveholders' expeditions. Forming stable communities was then contingent on their ability to practice agriculture and to extract the resources the forest provided, which would, later on, provide access to cash through commercial exchanges as well. Finally, by the time they were moving to the lakes in the mid-Trombetas, their navigational skills made them the best available pilots and guides of the Lower Amazon Rivers—a fact they might lament in the future. Command over natural landscapes not only permitted flight—it permitted autonomy and control over the return to life under Brazilian law, enabling the maroons to construct alternative livelihoods that subverted the rationale of slavery.

The relevance of this story is not restricted to Amazonia, as the insurgent geography of the Lower Amazon transformed it into one of the most important "black spaces" of northern Brazil.[65] During the 1995 celebrations of the 300th anniversary of maroon hero Zumbi dos Palmares's death held in Brasilia, the Trombetas community of Boa Vista became the first quilombo descendant community to receive a collective land grant from the federal government. The dam that earlier in the decade was slated to be built, prompting the maroons to write down the story of the Big Snake, never materialized due to the insufficiently high prices of bauxite in international markets during those years and to other factors as well. Currently, 361,825 hectares (about 0.9 million acres) have been granted to the twenty-five black communities living in the area, and another fifteen are waiting for the demarcation and titling of 332,654 hectares, or about 822,000 acres.[66] Like the Jamaican maroons, the Saramaka's Suriname River, or the seventeenth-century maroon settlement of Palmares, the Trombetas River is a territory in the Americas where enslaved Africans were able to sustain a protracted and frontal resistance to chattel slavery and, more broadly, to the colonial project of economic, social, and environmental domination.

Condensing visions of the future and understandings of the past, narratives like the Big Snake continued to be key to struggles over land and citizenship after slavery was abolished. When the viability of their communities was compromised by the advances of commercial houses, these narratives would become part of the shared understandings of community and citizenship held by the mocambeiros. As such, they would be used along with other discursive resources to oppose the threat of communal destruction and to enable the mocambeiros to claim their rights as Brazilian citizens.

CHAPTER THREE

I Do Not Buy My Freedom
Because I Am Not a Fool

Environmental Creolization and the Erection of
Communities in the Senzalas, 1850–1888

In 1848, British naturalist Henry Walter Bates went to Amazonia to study the biology and the natural history of the local flora and fauna. An astute observer of all things both natural and human-made, during his first years at the port city of Belém Bates visited a series of rural properties owned by the American and British residents of Pará. One such place was Maguari, a rice mill located about twelve miles north of the city that belonged to a Massachusetts merchant known as Mr. Upton. There Bates made the most unexpected finding: not an exotic carnivorous plant or a new species of butterfly—although he would eventually discover and study hundreds of them—but a young enslaved man called Hilario who, after meeting some English and American visitors in the past, "had showed his appreciation of their favour by picking a few words of English." Bates described Hilario as a "favourite young negro slave" and nicknamed him Larry. This polyglot man was an assistant and guide to foreign visitors in their excursions to the mill and its surroundings, paddling them to hunting spots and shooting alligators and waterfowl as he explained the secrets of the forest. Larry enumerated to Mr. Bates "the Indian names" and "the properties of a number of the forest trees," displaying his expertise regarding local flora and fauna. According to the naturalist, the slave "took an interest in our pursuit," and eventually became a valuable companion to Bates as he conducted his zoological and botanical activities.[1]

When he did not serve as a guide or hunter for visitors, Larry worked as a supervisor or accountant for the mill, "marking the sacks [of rice], and, being paid a price for all above a certain number, he earned regularly between two and three dollars a-week." This was a much more desirable job than being a field hand and denoted that Larry had achieved a position of prestige in Maguari. It was not only less strenuous than working in the field but also gave him access to cash and thus to better living conditions—perhaps even to freedom, if he could eventually save enough money to purchase it. When the American entomologist William H. Edwards met Larry some years later, he suggested that the slave "was in a fair way to be a freeman." Larry, however, "dispersed all such notions by the sententious reply 'I do not buy my free-dom, because I am not a fool.' He had a good master, he had a wife," explained Edwards, "and he did not have care or trouble. Thus he was contented."[2]

It is obvious that Larry's boastful exclamation contains a degree of decep-tion and performance and that Edwards was offering a correspondingly ro-manticized portrait of slavery, as foreign visitors in Brazil often did.[3] But underneath the surface, Larry's words speak of how enslaved people could rely on environmental creolization to alter the terms of daily life under slavery. By environmental creolization I mean the process of gaining familiarity with the opportunities and constraints of local environments, a first step toward suc-cessfully practicing agriculture, forest collection, hunting, fishing, and the trade in forest items.[4] While we do not know for certain why Larry attained a posi-tion of privilege, it is clear that his skills as a hunter and guide in the wilder-ness reinforced his status. By providing Larry greater access to broader social circles, to remunerated activities, and thus to the means to form a family, his mastery of the forest's ecological characteristics ultimately led to better living conditions and wider margins of autonomy inside of slavery. As I show in this chapter, Larry's experience epitomizes that of many other Paraense slaves who during the last decades of slavery developed an advanced mastery of the for-est's agro-ecological characteristics, acquired a specialized occupation related to it, and improved their living standards accordingly.

The first part of the chapter analyzes the different paths by which slaves acquired environmental knowledge: importing skills and strategies from equatorial Africa, acquiring knowledge when doing agricultural work in plan-tations and farms, and maintaining interethnic contacts with Indians.[5] As they learned the ways of local peasants, the enslaved gradually built entire parallel economies with vigorous ties to the expanding network of commercial houses that existed in late nineteenth-century Amazonia. Instead of using the traditional term for the economic activities that the slaves practiced for themselves—an "internal economy of slavery"—I conceptualize them as an economy that ran parallel to that of their masters, given the size and com-plexity of the commercial networks in which the slaves participated. Indeed, they tapped into an intricate system of itinerant merchants and commercial houses that spanned the entire Amazon basin and that grew even thicker as rubber and other exports skyrocketed in the second half of the nineteenth century.

Because the process of environmental creolization was both an individual and collective one, the chapter then turns to the formation of families inside of the *senzalas*, or slave quarters. Being part of a family brought greater possibilities for diversifying one's diet through gardening, increased access to cash, and a better quality of life. Families and parallel economies, then, fed back on each other, paving the way for the gradual emergence of extended families and of slave crews that sometimes resembled extended kinship groups with a vigorous collective culture—or, simply put, communities. The passage of the Rio Branco Law in 1871, which outlawed the separation of slave families after the slaveowner's death, and the masters' own agenda of minimizing disturbances in their slave crews were also pivotal to the formation of slave families. But the flexibility that the institution of slavery acquired over time also had negative consequences. First, the access of the enslaved to jobs such as artisans or extractors of forest items, which involved a degree of mobility, seems to have created a hierarchy of income and autonomy between enslaved men and women, deepening the gendered division of labor in the *senzalas*. Second, the slaveholders' adaptation to the ascendancy of rubber as Pará's primary export enabled Amazonian slavery to survive until May 13, 1888, when Princess Isabel finally signed the Golden Law. Just as the enslaved rendered slavery more flexible, the masters capitalized on the opportunities for prolonging the life of this nefarious institution.

This chapter also describes the process of community formation inside the slave quarters at the time of Amazonia's rubber boom. Rubber exports expanded throughout the second half of the nineteenth century and became the region's undisputed chief commodity after 1870, which had both a positive and negative impact on the prospects of Amazonia's black slaves. On the one hand, the boom in exports made it possible for them to expand their parallel economy. There were more commercial houses, more itinerant traders, more available occupations related to trade and transportation, more international visitors to serve, and a bigger demand for foodstuffs from regional markets—in sum, more opportunities to engage in activities other than planting cash crops for export—both within and beyond the masters' purview. On the other hand, that slaves could carry out such a broad scope of activities meant that the slaveowners' labor force was able to adapt to the changes brought by the rubber boom. Unlike the other northern and northeastern provinces of Brazil, Pará did not lose slaves to the coffee-producing ones of Rio de Janeiro, São Paulo, and Minas Gerais during the 1870s and 1880s. Instead, it absorbed more slaves than it let go. In Amazonia slavery was just like rubber: flexible and adaptable to multiple environmental conditions. That quality explained its durability.

Learning the Peasant Livelihood

While they retained significant parts of their African background, on their arrival in the New World African slaves started exchanging parts of their culture with European and Indigenous peoples. These "processes of both fusion and segmentation" have traditionally been studied in the realms of religion, folklore, popular celebrations, kinship relations, food systems, and many other areas of culture. Interactions between humans and nature, on the other hand, have received less attention. And while in the last years some scholars have looked at the "creole ecologies" that emerged out of the arrival of Africans to the Americas, the concept of creolization has not been applied systematically to the study of relationships between humanity and the natural world.[6]

In the context of Amazonian plantations and ranches, the daily agricultural tasks that the enslaved had to perform represented the first step in their gradual process of adapting to life in the tropical forest. As the rubber economy took off in the second half of the 1800s and surpassed all other exports by 1870, Paraense plantations diversified their produce and focused increasingly on the production of food crops and services to supply the expanding local markets. The most common food crop that Amazonia's slaves produced was bitter manioc, a tuber that local planters both consumed and sold to local markets. In 1849, Alfred R. Wallace described how the slaves living in the Jaguararí estate on the Mojú River were "engaged principally in cultivating mandiocca"; two years later John E. Warren observed "a number of old slaves engaged in making farina" at the sugar plantation of Caripé, "twenty miles away from Belém."[7] This "bush with starchy tubers" was planted normally in the uplands once or twice a year using the slash-and-burn technique, often in gardens "scattered about in the forest." Some gardens were also located on "islands in the middle of the river," explained Henry Walter Bates as he visited an *engenhoca* or rum distillery near Cametá.[8] Given the variety of manioc cultivars throughout the Amazon basin, it is not surprising to find manioc gardens even in the floodlands or inundated grounds of the Amazon basin, sharing the space with cash crops like cacao or sugarcane. The most frequently planted variety of the tuber had a high immunity to pests and could yield two harvests per year, making it a versatile, enduring, and high-yield foodstuff well adapted to Amazonian conditions.[9]

The only problem with this variety of manioc is that it contains prussic acid, which not only keeps most predators away but is poisonous for humans as well. To remove the poison and make this food more durable, the tuber is usually transformed into flour by a process that all rural Amazonian populations know very well: peeling the tubers, boiling them, squeezing them, grating them, and finally toasting the grated pulp in large, concave plates made of copper, called manioc ovens. The resulting flour and other subproducts of manioc constituted, as they still do today, the basis of the Amazon diet. They also represent "the main carbohydrate resource for about 800 million people" globally, making manioc the sixth major food crop in the world.[10]

Manioc flour was supplemented by a variety of other food crops, which slaves also cultivated as part of their daily tasks. One of them was rice, which had been an export item in the pre-Cabanagem decades, but by the 1840s was a foodstuff as well.[11] Henry Watler Bates, for example, saw "people of mixed white, Indian, and negro descent" selling "their little harvests of rice" to Scottish rice planter and merchant Archibald Campbell in the mid-1800s. Corn and beans were also a staple of the Amazonian diet. When French novelist Émile Carrey described a sugar plantation in Mojú in the 1850s, he noticed the presence of corn among other planted food crops, as did British and American explorers William Herndon, Alfred Wallace, William Edwards, Charles Brown, and William Lidstone.[12]

Paraense planters complemented the cultivation of these staples with colorful orchards of wild and planted fruit trees. At the earlier mentioned Mojú plantation, for example, Émile Carrey observed coconut, lemon, mango, cacao, pineapple, and orange trees. In his description of the Tauaú pottery factory, thirty miles south of Belém on the Acará River, Edwards reported in 1846 that the estate also included a "hill covered by orange and cocoa trees," and Alfred R. Wallace counted "a good many orange and mango trees" on a cattle farm in the island of Marajó in 1848. The presence of the trees did not surprise him, because some months before he had dined at the estate of a planter on the lower Tocantins River, "terminating [the meal] with fruit, principally pine-apples and oranges."[13]

Although the slaves' rations often included cured meat and fish from external suppliers, the planters often sent the captives to hunt game. Lowland pacas (*Cuniculus paca*), agoutis, *coatás* (white-bellied spider monkeys), tapirs, and crocodiles were among the most frequent prey in the forests and lakes around the plantations. In 1858, Bates observed how Major Gama, the Óbidos military commandant, sent "every week a negro hunter to shoot coatás." Edwards also saw the "traps set by the negroes for pacas and agoutis, or other small animals" near the plantation of Caripé, in Marajó, and French painter François Biard witnessed their special techniques of hunting snakes and their expertise in identifying "birds that I could not see in the foliage" of the forest. Fishing was also common. In the Lower Amazon Edwards visited a cacao plantation where "the master was absent, but the slaves had a number of fine tambaki [a large freshwater fish], and we purchased enough already roasted to last us to barra [i.e. Manaus]." In Marajó, Wallace had also seen slaves who "hunt alligators for oil" and who went on fishing trips using "a large drag-net fifty or sixty yards long." Initially they "had not very good fortune," Wallace wrote, "but soon [they] filled two half-bushel baskets with a great variety of fish, large and small, from which I selected a number of species to increase my collection."[14]

As the extractive economy grew in the second half of the nineteenth century, the slaves were also sent more frequently to collect rubber, Brazil nuts, and other forest items. In the Lower Amazon, for example, they were often put to collect turtle eggs on the shores of the area's rivers.[15] The extraction of rubber also figured among the activities performed by slaves: Edwards reported how in the Guamá River "what had been a cultivated plantation" with a number of slaves was now "growing up to forest, the Senhor having turned his attention to seringa." Near Baião, in the lower Tocantins River, the Jambuaçú plantation was combining the cultivation of cacao with the extraction of rubber.[16] Around 1870 rubber clearly gained momentum: a Belém newspaper advertised the sale of "a black slave 22 years old, working as a rubber tapper," and a number of slaveowners like Antônio José Henriques de Lima, José Joaquim Alves Picanço, and many others employed their African slaves in collecting seringa.[17] The *Hevea Brasiliensis*'s milk was becoming the engine of Pará's export economy.

In some respects, the productive activities that the enslaved carried out in Amazonia were not that different from what they had done in West Central Africa, the region of origin for most of the slaves arriving between 1800 and 1835.[18] By that time, manioc cultivation had spread across that part of Africa to the point of becoming, according to French explorer Paul du Chaillu, the "*aliment de prédilection*" for the populations of the Congo River.[19] In Africa, women planted the manioc tubers with the slash-and-burn technique after the men cleared the fields with the help of extended families and neighbors. Then the women produced processed foods such as manioc flour, *infundi* (manioc bread eaten alone or in stews), or *quiquanga* (manioc cakes).[20] Both the gendered division of labor and the recipes to process the manioc were almost identical to those of Amazonia. These foodstuffs were easy to store, durable, and tradable—which explained why manioc went from import to staple during the era of the slave trade, spreading gradually along West Central Africa's rivers.[21]

Similarly, the production and consumption of food crops as important to the local diet as rice and maize did not differ greatly between Africa and Amazonia. The slaves who arrived in Pará before 1800 had come predominantly

FIGURE 3 Slaves collecting turtle eggs, c. 1820. Spix and Martius, *Atlas zur Reise in Brasilien*, 27. Courtesy of Missouri Botannical Garden via Internet Archive.

from Guinea, an area with rich traditions of rice cultivation. As the number of free blacks and mulattos grew during the first half of the nineteenth century, many became involved in cultivating wild rice and selling it to the rice mills near Belém.[22] By the second half of the nineteenth century rice was planted over a large area in the region, as the presence of wild rice in the Upper Amazon attests. Although it originated in the New World, maize also had an African past that Paraense slaves may have known before their engagement with the crop in the New World. By the eighteenth century the crop had already spread through the savannahs of Angola, less so in tropical forests, and became a crop closely associated with the slave trade.[23]

The subsistence strategies employed by West Central African farmers in the tropical forests and savannahs of Congo and Angola were strikingly similar to those used in Amazonia. Bantu peasants in equatorial Africa, for example, combined agriculture, forest collection, and domestic orchards.[24] The main foodstuffs in the region were maize, beans, calabash, palms, and nuts, along with orchards of bananas, watermelons, and pineapples, just as in Amazonia, with some variations.[25] Collection of fruits, roots, nuts, ants, larvae, mollusks, fibers, medicinal plants, and drugs played a central role in the diet and the activities of both regions' inhabitants, frequently on a seasonal basis.[26] Amazonians and equatorial Africans were equally engaged in hunting fowl, small mammals, and crocodiles, and both were familiar with fishing techniques employing lances, nets, and poisonous plants to stun and capture fishes in rivers and lakes.[27] Last, but not least, rubber extraction was also known to equatorial Africans.[28] While by the second half of the nineteenth century only about 7 percent of slaves had been born in Africa, knowledge of rubber extraction acquired before the Middle Passage was useful for those who arrived in Amazonia during the first half of the century, when the milk of the *hevea* tree was used to fabricate a number of manufactured items, such as waterproof shoes and clothes.

Yet those who did not bring this knowledge from Africa could learn it from Indian and mixed-race individuals locally.[29] Foreign observers during the eighteenth and nineteenth centuries reported that Indians and African slaves worked together as laborers in the slaveholding properties of Amazonia and occasionally even caught a glimpse of the interethnic exchanges taking place between natives and slaves. Consider the "strange tableau" described by American diplomat John Esaias Warren in 1851:

> In the corner of the hut were a couple of negro women, seated on the ground engaged in basket-making; while a boy was cutting long strips from a species of cane used for this purpose. Various kinds of birds and skins of animals were hung around the cabin, together with ragged clothing, and bunches of fruit. One spectacle, however, which served to complete the picture, ... was that of an aged native, with whitened locks streaming down his shoulders, deliberately tearing to pieces, for the convenience of mastication, the body of a recently-roasted Guariba, or howling monkey.[30]

The "negro women" and the young man in this vignette were making "baskets," most likely *tipitis*, the long, braided cylinders used to squeeze the boiled manioc pulp before grating it—an omnipresent artifact in Amazon farms. Similar weaving techniques were employed to make hammocks with cotton and other fibers, a household technology frequently used by slaves. If we consider in addition the "birds and skins of animals," the "bunches of fruit" hanging from the walls (bananas, *açaí* berries), and the roasted *guariba* monkey the Indian man was eating, this scene speaks of a rich process of knowledge sharing about what—among the vast repository of products of the tropical forest—could be grown, collected, processed, built, conserved, fabricated, hunted, fished, distilled, bought, and sold. Scenes like these explain why in 1846 entomologist William Edwards manifested no surprise at all when he saw, near the neighboring state of Amazonas, "two free negroes who had been admitted to the rights of tribeship" in an unspecified Indian tribe.[31]

Indeed there was a long history of Indian-slave interaction. In the eighteenth century, most of the largest plantations were owned by the Catholic orders and included significant numbers of enslaved Africans and Indians toiling together.[32] In the next century the presence of Indians in local plantations clearly diminished, but continued to be visible along the ever-expanding frontiers of interior Amazonia. On the Capim River, for example, Wallace visited Senhor Calixto's São José rice and sugarcane plantation, where he saw "about fifty slaves of all ages, and about as many Indians." Nineteenth-century visitors to Paraense plantations and livestock farms often saw "natives and slaves employed" together—especially in areas farther from Belém, where Indians were more numerous. Mundurucú Indians, for example, were frequently found performing varied tasks on cacao and tobacco plantations, on cattle ranches, and in the streets of the Lower Amazon cities.[33]

Local planters, in fact, preferred Indians and African slaves to live and work together. Because of the labor shortage in the region, landowners tried to secure whatever workers they could find—probably by subjecting the Indians to forms of concealed servitude such as debt bondage, given that their enslavement was illegal.[34] A planter named Calixto explained how "by having slaves and Indians working together he was enabled to get more work out of the latter than by any other system." British sugar planter Mr. Culloch agreed. He explained to botanist Richard Spruce in 1850 that while "he employed several native handicraftsmen ... the only workmen on whom he could rely were four slaves of [his Brazilian associate] Henrique's." The former Confederates who settled in Santarém after the U.S. Civil War had a similar opinion: Mr. Rhome, the only one among them whose plantation apparently succeeded, did so because he "found a rich and enterprising partner, with thirty or forty slaves to do his work." These are prime examples of race management; that is, the "claims to know the fitness of certain peoples for certain jobs and to develop 'lower' races by slotting them into, and disciplining them through, certain types of labour."[35] For Amazonian landowners, forcing Indians to work alongside African slaves made the natives better workers.

From Learning to Practicing: The Slaves' Parallel Economy

By the mid-nineteenth century, the enslaved were putting the agro-ecological knowledge that they imported from Africa and learned from Indians in the service of gaining autonomy and self-control over their daily activities. The first place where they did so was their provision grounds, the small land patches provided by the masters so that the slaves could produce their own food on their weekly day off—usually a Sunday. In those grounds the enslaved found a measure of autonomy, a chance to experiment, access to cash, and, gradually, knowledge of what being a free peasant meant. Classic studies on provision grounds in the United States, the Caribbean, and other Brazilian regions have also shown this process.[36]

Amazonian planters, like those throughout the New World, fed their slave crews through the purchase of food, the cultivation of food crops, and the grant of provision grounds. "Upon Saturday afternoon," described Edwards after visiting the Tauaú pottery factory in 1846, "all the blacks collected around the store-room to receive their rations of fish and farinha for the ensuing week." Many of them "had fowls and small cultivated patches, and from these sources, as well as from wood and river, obtained much of their support," added Edwards. Wallace observed in a large cattle estate in Marajó how the labor force, comprising twenty slaves and a similar number of free laborers, was "allowed farinha only; but they can cultivate Indian corn and vegetables for themselves, and have powder and shot given them for hunting, so that they do not fare so badly." On "Sunday" they enjoyed a day off "for working in their gardens, hunting, or idleness, as they choose." In addition, "many of them keep fowls and ducks, which they sell, to buy any little luxuries they may require, and they often go fishing to supply the house, when they have a share for themselves."[37] Using whatever time they could scratch from their working hours, Paraense slaves created a space to experiment with subsistence strategies, gaining some autonomy and training themselves to become free men and women someday.

While these economic activities are usually considered *internal* to the economy of slavery,[38] the slaves continually tried to transform them into an entire *parallel* system. With or without their masters' acquiescence, the captives gradually transformed small-scale agriculture and forest extraction into a means to start trading with the network of itinerant merchants or regatões that spanned the Amazon basin. This was part of the process of environmental creolization, as trade with the regatões was integral to established agro-ecological practices in Amazonia. These commercial exchanges also had the potential to improve the slaves' living conditions and to sustain their families and communities, gradually eroding the domination inherent in slavery and in some cases leading to freedom through routes blazed both individual and collectively.

Roaming the region's rivers, the mobile regatões constituted a promising resource for the enslaved. The first contacts with them began as part of the captives' work routines. For instance, the young slave José, 10 years old in 1860, and Joana and Joaquim, both in their thirties, were frequently sent by their master Marcella Maria de Santa Anna to a nearby commercial house to buy groceries and supplies for the small farm she had in Abaeté, a sugar-producing municipality in the lower Tocantins River. Ms. Santa Anna also paid back her debts by renting the work of the slaves Joaquim and Benedito by the day, although we do not know exactly what activity they performed. Felipa de Jesús Furtado's slaves Leandro and Thomé, 21 and 46 years old in 1873, respectively, were also sent by their master very often to buy *pirarucú* fish and fresh meat at a merchant store on the Capim River.[39]

But only one step separated trading for their masters from trading for themselves. In 1865, when Portuguese merchant José Maria de Andrade died in São Domingos do Capim, he listed among his debtors the slave Anacleto, who owed him 37 milréis, and Antônio dos Santos's slave Manoel do Carmo, who had a debt of 10,500 réis. The slaves Eleuterio, Jacintho, João, Isidora, Manoel José, Marcolino, Manoel, and Nazario also had open accounts with Andrade for variable sums. In Santarém, a number of slaves were also included among José Roiz dos Santos Almeida's debtors when his commercial house had to declare bankruptcy in 1870; their debts ranged from 250 réis to almost 30 milréis. In the city of Alenquer, a man called Manoel, "Demetrio's slave," had contracted a debt of 11 milréis with Brasilino Caetano José da Costa, a local merchant, in 1879.[40] As the enslaved themselves practiced the commercial exchanges they carried out for their masters, they gradually learned the emancipatory potential of trade.

Naturally, the slaveowners were greatly perturbed by their slaves' trade activities; indeed, the slaves' clandestine trade was seen as an endemic problem in the Lower Amazon. "Can anybody doubt," lamented in 1850 a councilman from Óbidos, "that the regatões have caused this and other evils plaguing our formerly flourishing county?" In 1864, the provincial president complained, "The landowners who live in the interior suffer the activities of ruthless merchants [*vendilhões*], who live near the estates and trade with the slaves, thus encouraging theft." In response to these concerns, the state government on May 1, 1877, established a fine of "20,000 réis or five days of prison to those trading with slaves" without the masters' permission; five years later it instructed

rural police officers to prevent the gatherings of "cowboys and slaves" in commercial houses across the island of Marajó. But illegal trade persisted: in 1881
the Municipal Council called for the enforcement of fines against regatões for
"trading with slaves without their masters' permission."[41]

In addition to engaging in illegal relationships with itinerant merchants,
slaves like Larry, the forest guide discussed earlier, acquired specialized skills
as hunters and backwoodsmen, allowing them to freely extract forest items and
to become guides and rangers for foreign visitors in hunting expeditions—visits
that were becoming more frequent as the century drew to a close. Larry was
probably the only slave who was able to speak English, but many others worked
as professional hunters and rangers. That is the case of Juvenal, a slave descendant portrayed by writer Sylvia Helena Tocantins in her memoire *At the
Trunk of the Sapopema*, and of Luiz, a former slave born in Congo and brought
to Belém via Rio de Janeiro by Austrian naturalist Johann Natterer. Luiz could
eventually purchase his freedom with the money he saved while working as a
bird catcher for the naturalist.[42]

The rubber trade expanded these opportunities. By the 1870s, a growing
number of slaveholders increasingly turned to extracting the hevea milk, which
in some cases was already present on their property. Others purchased properties with rubber trees, allied with merchants, or invested in sectors that were
expanding due to the indirect impact of the rubber bonanza, such as transportation, urban services, real estate, or foodstuff production for the growing
population of the state. Take for example the Corrêa de Miranda, a dynasty of
powerful slaveholding sugar and cacao planters from the county of Igarapé-
Miri, the heart of rural slavery in Pará. While the dynasty's wealth came traditionally from cash crops grown by slaves, which they possessed by the hundreds,
by the 1870s individuals like Manoel João Corrêa de Miranda or Justo José
Corrêa de Miranda started investing heavily in rubber trails and on real estate
in Belém, respectively.[43] Manoel João also became the godfather of Valeriana
Castro, the daughter of powerful rubber exporter Francisco da Silva Castro—
an alliance that surely facilitated the conduct of business. As the rubber era
took off, traditional slaveholders faced the choice of either participating in
the new activities or lagging behind in access to wealth and power.[44]

In the lower Tocantins and the rivers flowing to Belém, blacksmiths,
carpenters, weavers, mechanics, and canoe makers and pilots were in great
demand by the 1870s, needed to service the booming fluvial traffic of ships
carrying rubber, produce, and manufactured goods. While it was dangerous,
working as a canoe pilot or rower could yield sizable cash rewards. Such was
the case of the slave Miguel, from Marajó, who made a staggering 60,000 réis

for working two months as a canoe pilot in 1882. Canoe making was common in a riverine environment like Amazonia, as revealed by the frequent appearance of entire sets of carpentry tools in postmortem inventories. When a dispute over the estate bequeathed by Antonio José Henriques de Lima to his son Raimundo José took place in the Acará River area in 1874, the heir complained to the judge that his father's slaves had been "very poorly paid" by the commercial houses that hired them. "The *carafuz* [mixture of Indian and black] Joaquim is a master carpenter and a caulker, as the [population of the] Acará and Pequeno Rivers know." The actual price for a day of the slave carpenter's work, of which his owner received a part, was the considerable sum of 2,600 reis, Raimundo José claimed. "I appeal to the people of the Acará to rebut this truth," he added, invoking his neighbors as witnesses, and making it evident that throughout the region Joaquim was indeed known as a prestigious carpenter.[45]

As the number of slaveholding properties relying on cash payments grew by the 1870s, slave parallel economies grew thicker. For the slaveowners, a task system with some payments for overwork was convenient because apparently it made it less problematic to employ free laborers alongside slaves, as happened in other slave regions of the Americas. For the captives, paid extra work provided yet another way to obtain a cash income and, eventually, even freedom.[46] Maria Barbara Gemaque Pereira, a livestock owner from Marajó, made numerous payments to her slaves in 1872, and so did the late Antonio José Henriques de Lima in an agricultural and rubber-producing property of the Acará River. The slaves Lourenço, José Joaquim, Manoel Jorge, and Roberto, who lived on the same plantation as the "famous" artisan Joaquim cited earlier, saved enough cash to purchase their freedom on their owner's death in 1874.[47]

As the masters became more interested in spending more time in the cities, some slaves were left in charge of managing the daily operations of plantations, ranches, and farms. British naturalists Charles B. Brown and William Lidstone visited a farm in the mid-1870s where "the owner lives at Monte Alegre, and leaves the place in charge of six negroes, five of whom are slaves." Near Santarém, at the Fazenda Santa Anna, they met another slave "who has the entire charge of the place for its non-resident proprietor—a dweller in Santarém."[48] Productive diversification, an expansion of wage labor, and a blooming urban economy that pulled rural elites away from their estates were the forces, generally welcomed by the enslaved, that altered the terms of rural slavery in Pará. Relying on expanding social and commercial networks that clearly surpassed the boundaries of their masters' estates, it is no exaggeration to argue that, more than being merely an internal system, the slaves' own economy eventually became parallel to that of their masters.

Family

As we saw at the beginning of this chapter, while the slave Larry invoked a privileged job and a good relation with his master as reasons why he did not want to become free, having created a family of his own was no less central to the autonomy he boasted. That Larry "had a wife" was one of the main reasons why he "was contented," according to entomologist William Edwards.[49] More information about the characteristics of Larry's family is unfortunately not available, but evidence about the slave families that lived in Pará's senzalas does indicate that, between 1850 and the abolition of slavery in 1888, the slaves gradually created multigenerational communities based on extended kinship, making this institution pivotal to their bid for autonomy and freedom.

In 1850, the slave labor force of Pará reached its demographic peak in the post-Cabanagem decades, numbering 33,542 individuals representing 20.33 percent of the state's population. As discussed in chapter 2, the number of adult males and females in local slave crews decreased because of the revolt, but this did not hinder the natural reproduction of the enslaved. Much to the contrary, the natural growth of the slave population that was already present accelerated because the revolt further balanced the slaves' gender ratio and removed ethnic barriers to the formation of families.[50] By the mid-nineteenth century, the demographic creolization of the Paraense slave population had consolidated.

During the 1850s, then, the slaves rapidly formed families and raised numerous offspring, as evinced by the size of the slave children population. As seen in graph 2, children younger than age 14 constitute the largest age group in the slave population: 38.2 percent of the total. The ratio between slaves ages up to 10 years old (children) and women aged 15 to 49 (potential mothers) was 1,113, a high value similar to that of other regions where the enslaved experienced sustained natural growth, such as Minas Gerais (980 for 1808, 925 for the 1850–1888 period).[51] While the existence of a large number of children is not by itself an unequivocal sign of family formation, it seems that Amazonian slaves opted for having large numbers of offspring, reproducing themselves quickly and staking a claim over their future on the land.

The kinship ties we can reconstruct using probate records suggest that nuclear and single-parent families tended to be the most common family types. It is not possible to determine which form predominated, because postmortem inventories routinely ignored fathers and registered only the ties between mothers and their children—and sometimes not even those. After the passage of the Free Womb Law in 1871 mandated the court system to register the slaves' demographic data more carefully, kinship ties started appearing more often in probate records.[52] That legal marriages among Paraense slaves were rare also makes it difficult to establish how stable formal marital unions actually were. According to the 1872 census, only 6.17 percent of all Paraense slaves were married, and another 2.16 percent were widows and widowers. Marriage rates among the free population were also low: 22.32 percent were married, and another 4.4 percent were widows. The national average was between 20–33 percent higher, both for the slaves and the free population.[53] In Amazonia, the lower marriage rate was probably due to "the lack of parish priests in the interior," as both visitors and locals often lamented.[54] In sum, while slave fathers like Larry were a real figure in Amazonia, the actual contours of their role in slave families will have to remain sketchy for now.

In Pará single-parent and nuclear slave families tended to have, on average, 2.1 children—and slightly more in large slave crews.[55] This number reflects those registered as sons and daughters at the time their owner passed away, but the real number of children per family was probably higher, given that sometimes bonds of parenthood were only registered when the sons and daughters were minors. This average also reflects the number of surviving children. Child mortality rates among slaves in nineteenth-century Brazil were high: about one-third of all slaves died in their first year; in Bahia, 47.2 percent of slaves died before their eleventh birthday.[56] Finally, the families formed by the slaves often faced the possibility of being separated after the death of the slaveowner, when slave crews were divided. In sum, the average of a bit more than two children reflects the success of the enslaved in enduring all the forces fragmenting their families, from poor living conditions to Brazilian inheritance law.

According to probate records, in their daily lives 49.6 percent of all Paraense slaves shared residential and work spaces with at least one first-degree family member. However, if we take into consideration that fathers were mostly not recorded, as mentioned earlier, this seems like a conservative figure. If we assume that each mother shared the household with a father, then the percentage of slaves living with at least one relative jumps to 65.12 percent, a figure that is probably closer to reality.[57] Nor did probate records register forms of fictive kinship such as the traditional Brazilian godfather or the African *malungu*, who could play as relevant a role in the family as blood kin.[58] But even ignoring fictive kinship, we can argue that about two-thirds of Amazonian slaves normally lived and worked on the same workplace as at least one of their first-degree relatives, a percentage considerably larger than the Brazilian average during the 1870s or than the average of the lower Tocantins area of Pará itself in the decades before the Cabanagem revolt.[59] Thus, by the 1850s most Paraense slaves shared their daily experiences with their kin, be they spouses, siblings, or grandchildren.

By the time the Brazilian Empire recognized the importance of slave families with the passage of the 1871 Rio Branco Law, which established the freedom of children born from slave mothers, multigenerational slave families residing in the same property had become a common occurrence in rural Pará. Whereas in the 1850s, there is evidence of only five such families in a sample of 635 individuals in forty-two postmortem inventories (although their real number was probably higher), in the 1870s no less than seventeen slave families with three different generations lived in the same property, out of a sample of 726 enslaved individuals living in fifty-nine slaveholding properties. The average number of children per slave family had increased slightly to 2.17, and the proportion of slaves living with a relative of any degree was 50.69 percent—or 66.12 percent if we add a father for each of the nuclear families counted. Slaveholding estates with thirty slaves or more tended to host more family units: 52.72 percent of the slaves lived there with a relative, or more than 70 percent in a hypothetical count including fathers.

From the slaves' standpoint, finding a suitable mate and starting a nuclear family provided benefits at different levels, starting with the emotional "solace and support" of having a partner and descendants.[60] The judicial actions taken by hundreds of Paraense slaves in the aftermath of the 1871 Free Womb Law to maintain the integrity of their family groups attest to the importance of family for them. A slave woman called Francisca, for example, explained to a Belém court in 1874 that she was unjustly "subject to the captivity" not of her own body, but "of a daughter minor of age named Joanna," taken to Europe by her alleged owner in flagrant violation of her legal freedom.[61] "Casemiro Antonio Porfirio, his mother Raimunda Maria de Santanna," and six other siblings, in another instance, requested that the judge grant them their manumission letters in 1875, at the municipality of Breves.[62] They were trying to attain freedom collectively, as a family. In other words, in Pará, as in Rio de Janeiro, "barely if ever did the goal of manumission appear as a purely individual one. In reality, the transit [from slavery to freedom] was only complete when the whole [family] group lost its links to captivity."[63]

From a more pragmatic perspective, the family group also represented the main productive unit and the most immediate source of labor to make possible the gradual adoption of a peasant way of life.[64] In this sense, slave families developed a gendered division of labor following the common Amazonian model: men would engage with activities that involved spatial mobility, such as hunting or seasonal extractivism, and women would tend to gardens and carry out other potentially productive household activities, such as the production of handcrafts. This gender-split household economy appears in the tableau with the "negro women, seated on the ground engaged in basket-making" discussed earlier.[65]

The slaves found an unlikely ally in their efforts to form families: the slaveholders. The chronic scarcity of labor that characterized the Amazonian economy in the second half of the nineteenth century motivated planters to keep their slave crews together. Thus, aware that breaking up a slave family could increase the risk of slave flight, when a slaveowner passed away his or her heirs frequently tried to keep slave families together, to the extent that Brazilian inheritance law permitted it.[66] In addition, keeping slave families together across the generations also generated decades-long attachments between masters and slaves, which in turn reinforced the construction of symbolic patriarchal ties between both classes. Sanctioning by law the de facto situation, the 1871 Rio Branco Law prohibited the separation of slave families on the occasion of a slaveholder's decease, institutionalizing this trend toward safekeeping slave families.[67]

As the slaveholding elite of Pará diversified its activities in the wake of the rubber boom and started investing more heavily in other productive sectors, rural estates were sold or mortgaged with their entire slave crews attached as one more asset, along with the machinery, cultivated grounds, orchards, and cattle. In 1872, the Murutucú plantation contained "a steam-powered sugar mill to fabricate sugar and cachaça, hydraulic sawmill, barracks, harvesting tools, coils, pipes, diverse utensils, 10 head of cattle, and fifty slaves;" it was valued at 45.5 contos. Not far from there, in Abaeté, cacao planter and merchant José Ferreira Bello mortgaged his Arumanduba River cacao estate in 1878, including "the agricultural tools and belongings of this property, and the [six] slaves as an integral part of it." In Vigia, Agostinho José de Almeida sold the Santo Antônio da Campina sugar plantation to the baron of Guajará in 1874 for 35 contos, with its fifty-eight slaves included in the price. Attaching the slave crews as an asset seemed to be one of the few ways to maintain a stable labor force in large agricultural estates.[68]

Given this confluence of slave and master priorities, by the 1870s the crews of some large properties had come to resemble extended kinship groups more than a labor force. Maria Thereza Maia e Miranda's Livramento plantation, for example, employed a labor force of twenty-three slaves in 1876. Victor, Gabriel, Pedro, and Eloi, aged between 23 and 29, were brothers and lived with 51-year-old Mónica, a widow. They were born and raised there. Cândida, a 25-year-old unmarried washerwoman, lived with her sons Aleixo and Maria das Neves, ages 4 and 7. In addition to these families, the slaves Romão (39), Francisco (41), and Felix (51), brothers and sons of an unknown mother, and siblings Leocadio (35), Hilaria (23), and Luiza (18), all lived in the same plantation. More than half the slaves on the property, in other words, were part of three family groups. The São José sugar mill, in Abaeté, hosted at least seven families among its forty-seven slaves, including Marcella's five siblings, Maria de Nazareth's four siblings, Lina's four siblings, and another four from a different Maria, in addition to Martinha, Maria Sophia, and Jacintha, who each had one child. And these are only the kinship ties we know—the unknown fathers and grandparents probably played a role in the family as well.[69] In sum, even before the passage of the Rio Branco Law, Paraense slaves' strategies of family formation, sometimes aided by the masters' own agenda, had succeeded in weaving families into the fabric of Amazonian slavery.

Paradoxically, as the density of slave families peaked in the early 1870s, the family structure also started to come undone. As shown by the age pyramid of Paraense slaves for that decade (see graph 4), the Rio Branco Law, by emancipating the children of enslaved women born after 1871, undermined the presence of slave families in the senzalas for the rest of the decade. The children to adult women ratio fell from 1,113 in the 1850s to 678, evincing the absence of *ingenuos* (children of enslaved women born free after the 1871 Free Womb Law) from the rural estates' slave crews. In the 1850s, the dependency ratio, or the relation between nonproductive (ages 0–14 and 50+) to productive (age 15 to 49) slaves, was 1.01, which meant that the number of elders and children surpassed that of adults; in contrast, the 1870s the dependency ratio was 0.78, meaning that there were more slaves in prime productive age than either elders or children. By emancipating children, the 1871 Free Womb Law inadvertently brought the demographic profile of Pará's slaves closer to the slave masters' ideal.

This does not mean, however, that the children born free were necessarily separated from their parents. By the time sugar and cacao planter Antônio Francisco Correa Caripuna died in 1877, for example, eight of his forty-seven slaves had mothered a total of fourteen ingenuo children who were counted in their owner's postmortem inventory, despite being legally free. This was an unequivocal sign that they were still living together—and that those emancipated by the Rio Branco Law were at risk of being treated as de facto slaves.[70]

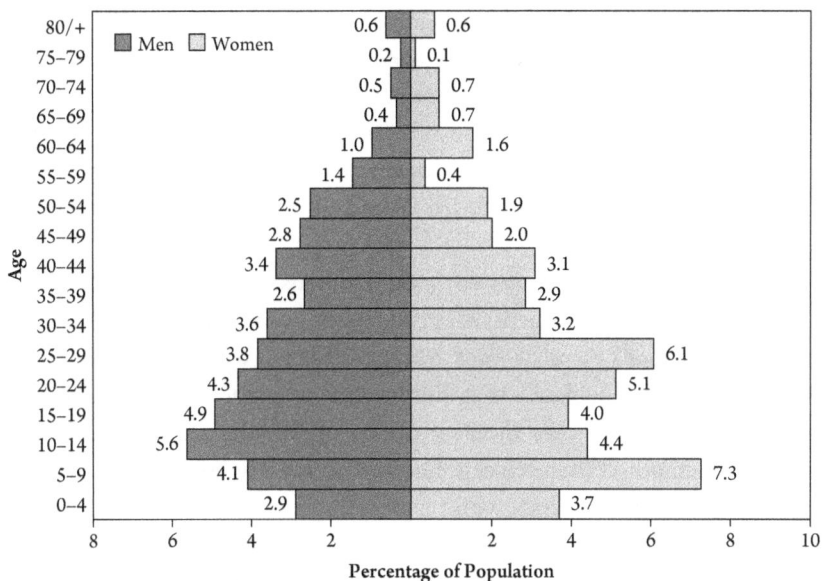

GRAPH 4 Age pyramid of rural slaves in Pará, 1870–1879. Source: Subsample of seventy postmortem inventories from 1870–1879, totaling 772 individuals with a known age (approximately 2.8 percent of the total slave population in that decade). Funds CBA, AFS, APEP-IM, APEP-1V, APEP-2V, 5A-CR, CMA, and ITERPA-LPTV.

As demonstrated by the actions of Lucrecia, Ursulina, and Maria Theophila, Caripuna's enslaved mothers, the liberation of slave wombs did not stop the formation of slave families—it simply displaced the newer generations to the realm of legal freedom.[71]

The Legacies and the Problems of Larry's Strategies

While the parallel economies and the multigenerational families discussed in this chapter allowed the enslaved to build communities and approach a peasant lifestyle, they sometimes had unexpected negative consequences as well. The emergence of a slave parallel economy and the increasing access to specialized occupations, for example, could lead the slaves to reproduce "among themselves the unequal reciprocities" of gender that so profoundly shaped Brazilian society.[72] Of a sample of 2,138 slaves appearing in rural post-mortem inventories between 1850 and 1880, 137 or more than 6 percent of them were listed as having an occupation other than field hand. For enslaved men the listed jobs were masons, carpenters, blacksmiths, shoemakers, and other artisanal and technical ones (including twenty-two listed as officers of their trades). The women were cooks, laundresses, seamstresses, domestics, and one "doorwoman," totaling only thirty-six cases or a meager 1.5 percent of the total.[73] While the jobs for male slaves likely yielded cash for extra work and a higher degree of geographic mobility, jobs for women were restricted to the farm and the household and apparently provided little monetary compensation.

However, that enslaved women worked in household activities, including gardening and small-scale artisanal production, was not necessarily limiting, given that the men's earnings were most likely shared with their family members. There is also evidence that, despite this unequal access to paid jobs, rural enslaved women used their control over gardening and small-scale agriculture to empower themselves as heads of their households and to conduct trade with the regatões. Thus, while visiting the lower Tocantins River area in the early 1850s, naturalist Alfred R. Wallace met "a Negro woman" who was in charge of the farm where she lived, "because the master was not at home." His partner Bates also related the case of "an old negress named Florinda" who had become "the feitora or manageress" of an agricultural estate some miles away from Belém.[74] By the late 1800s and early 1900s, the role played by some old black women as leaders of their communities in the coastal city of Vigia led to the emergence of the *tias do carimbó*, elderly women recognized as community leaders and organizers of carimbó performances, a social occasion for all Vigienses.[75]

Nonetheless, the unequal access to cash between male and female slaves should not be taken lightly. In spite of the sharing of resources within the slave household, it is highly likely that this sharp gendered distinction in access to cash among rural slaves severely limited the enslaved women's chances to move up the social ladder. While in the urban spaces of Amazonia enslaved women masterfully used the streets to access cash and broader social networks,[76] in the countryside their sphere of influence was usually circumscribed to the household and the community, and they faced severe restrictions in accessing cash.

Similarly the concessions wrestled from the masters had mixed results: they increased the Paraense slaves' personal autonomy, but also ended up reinforcing the system of forced labor by making it more durable and better adapted to the new economic realities of the rubber boom.[77] Table 1 shows how, contrary to the conventional wisdom that Paraense slaveholders were quickly replaced by rubber barons, a number of masters held onto their slave crews until well into the 1870s. While the enslaved rapidly shrank as a percentage of

TABLE 1 Enslaved population in Pará and Belém, 1778–1888

Year	Pará			Belém		
	Total pop	Slaves	%	Total pop	Slaves	%
1778	54,914	12,067	21.97%	5,276	2,733	51.80%
1797	70,573	19,586	27.75%			
1823	128,127	28,051	21.89%	12,471	5,719	45.86%
1833	149,854	29,977	20.00%			
1848	164,949	33,542	20.33%	16,092	5,085	31.60%
1856	196,967	32,961	16.73%			
1862	215,923	30,623	14.18%			
1872	275,237	27,458	9.98%	34,464	5,087	14.76%
1884	274,883	20,849	7.58%			
1888	280,676	10,535	3.75%	40,000	2,196	5.49%

Sources: Anderson, "Following Curupira," 101; Bezerra Neto, *Escravidão negra no Grão-Pará*, 226; Pará, *Relatorio ... 1852*, 88; Roller, "Colonial Routes," 288.

the total population between 1850 and 1888, this was due to the enormous expansion of the non-enslaved population, not to a sudden drop in the total number of captives. The slave population stayed at approximately 30,000 individuals until the Rio Branco Law started to undermine it in the mid-1870s, which speaks of the elite's capacity to preserve its slave labor force until very late by making concessions to the captives.

Moreover, unlike the slaveholding elites of northeastern Brazil, during the second half of the nineteenth century Pará's masters did not lose slaves to Rio de Janeiro, São Paulo, or Minas Gerais, the coffee-producing areas of southeastern Brazil. Whereas between 1872 and 1885 the northeastern provinces lost captives to the southeast by the thousands every year, the state of Pará exported annually between 4,500 and 5,000 slaves, but imported a bit more than 5,500 as well. Not only did Pará retain its slave labor but also, "albeit secondary or residual, it became an intra-regional market attracting captive workers from the Northeastern provinces" through the port of Belém.[78] The surplus of slaves in Pará was not huge, but it does call for a revision of the vast majority of the literature about the interprovincial slave trade in Brazil, which often fails to recognize that, unlike all the other northern and northeastern states of Brazil,

Pará attracted more slaves than it expelled in the last decades before emancipation.[79]

The micro-social processes of environmental creolization discussed in this chapter, then, may have inadvertently served the interests of Pará's slaveholders too. The formation of slave families probably made life under slavery more palatable for the slaves themselves, as the example of the slave Larry shows. The expansion of transportation and trade networks increased the chances that the captives could work as carpenters, mechanics, stevedores, ship pilots, rubber tappers, or even cashiers, opportunities that they embraced as a way of accessing cash and saving money for their freedom and for their dependents. However, while all these developments allowed the enslaved to gain autonomy, they also made slave work more flexible and negotiable, prolonging its usefulness.

Yet, one should not lose sight of the fact that, by acquiring an advanced knowledge of local environments, the enslaved confronted, and ultimately reversed, the rationale of slavery in Amazonia. If the essence of chattel slavery was bringing an "exotic workforce" to the New World, a source of "mechanical energy" mutilated of its social ties and its agro-environmental knowledge, then reconstructing such knowledge represented an attack on, and ultimately the defeat of, the masters' rationale.[80] Moreover, the slaves' environmental creolization underpinned other forms of slave resistance, such as the erection of a vigorous parallel economy, the formation of families, and the use of market forces to eventually purchase manumission. Against dehumanization, exploitation, and deprivation of social bonds, the enslaved Africans built links of kin and place that were profoundly mediated by the environments where they lived and worked, thus carrying out a true "remapping of everyday life" that gradually allowed families and communities to enter the ranks of the free peasantry even before abolition came in 1888.[81]

By the time the slave Larry became an adult in the 1850s, he had a sophisticated knowledge of the Amazonian biota's economic possibilities, and he had formed a family that very likely included children; hence his boastful exclamations in front of foreign visitors. In his apparent rejection of legal freedom he was actually making reference to the substantial gains in daily life that he, his family, and many other slaves were achieving in those years as a result of an advanced environmental creolization. By the 1870s, the changes brought by the rise of the rubber economy only allowed them to advance further in the creation of parallel economies and multigenerational families, leaving behind endogamic marriages inside of African ethnicities, and allowing them to build communities inside the senzalas.[82] The ascent of rubber made the opportunities that Larry had enjoyed available to more people.

Brazilian historians have argued that two of the roles played by enslaved communities were to "define the emerging Afro-Brazilian culture" and to "socialize children to these beliefs and behavior."[83] However, the narrative and material dimensions of the slaves' natural landscapes were also a fundamental part of that socialization and therefore another important vehicle for the nascent Afro-Brazilian culture. As we will see next, such narratives and practices survived the end of slavery, shaping the trajectories of Afro-Paraenses in the post-emancipation decades.

CHAPTER FOUR

Working Almost as Slaves?

The Post-Abolition Brazil Nut Trade, 1888–1930

In the collective memories of the Lower Amazon maroons, the decades after emancipation are remembered as a period when "the people were oppressed" by Brazil nut merchants, who "enslaved" the blacks of the region. Emphasizing the continuity with the time of captivity, use of the metaphor of a continued slavery has become commonplace to describe how the commercial houses of the Lower Amazon acquired concessions for the public lands containing Brazil nuts and subjected the mocambeiros, as the maroon descendants were called even after the end of slavery, to debt bondage.[1] Scholars often agree with the idea that these decades represented a "new version of slavery" and that, in the words of maroon descendants, "we worked almost all slaves," despite the end of formal slavery in 1888.[2]

However, a number of individuals also relate memories of Brazil nut merchants who "helped the people" of the river, who "gave goods for the saint patron's parties," and who acted, in the words of a maroon descendant of the Trombetas River, as "fathers of the people."[3] The merchants sent food for special occasions, were godfathers for local children, and provided varied goods on credit to the maroon descendants. Naturally, the first story of this kind I heard surprised me, given the antithetical image of the period cast by the new slavery narrative. The challenge in this chapter, therefore, is to reconcile these conflicting representations of Brazil nut merchants and to discern what they tell us about the maroon descendants in the decades after slavery.

To reconcile these perspectives I will argue that these two conflicting stories reflect two different spheres in the relationships between black peasants and Brazil nut merchants. While the first one was characterized by domination, a few individuals successfully accommodated to, and even collaborated with, the newly arrived commercial houses. Those who spoke positively of the merchants were simply juxtaposing those two facets: the ruthless exploiter and the merciful patron, the boss who cheated the nut collectors and the generous financer of community activities. In both spheres, however, Afro-descendant forest specialists and explorers were fundamental to the merchants' penetration into a world where the mocambeiros had hitherto ruled. Just as in the past, knowledge of local environments turned out to be pivotal to determining who would gain the battle for the Brazil nuts. In the end, the loss of autonomy and lower quality of life in the 1910s and 1920s shaped the maroon descendants' social memory for the rest of the twentieth century, filling it with narratives of poverty, dispossession, and the speech figure of the "new slavery."

Post-Emancipation Mocambeiros:
Society, Culture, and Economy

The abolition of slavery on May 13, 1888, meant technically that the maroon descendants of the Lower Amazon could now enjoy unrestricted spatial mobility along the area's rivers.[4] In comparison with the pattern of settlement employed in the past, determined by the necessity to stick together and defend themselves, they could now disperse throughout the area's rivers, as noted by travelers and explorers. When French explorer Otille Coudreau visited the Trombetas River, for example, she saw "fifty of these mocambeiros" still living in the village of Colônia that the missionary Father Carmelo de Mazzarino had founded decades before, but she also noticed the "relatively numerous shacks" with cacao trees around their houses on both shores of the river. Nine years later, Austrian entomologist and botanist Adolpho Ducke observed how the "descendants of the ancient 'mucambo' . . . are spread further downriver" from the settlement of Colônia, "at [Lakes] Arrozal and Tapaginha." At Lake Abuí he also counted a "few residents, the remnants of the *mucambeiros* and their descendants, today perhaps about 30 people." In the 1920s scientists Gastão Cruls and Peter Paul Hilbert continued to notice the presence of mocambeiros all over the Trombetas and Cuminá and to identify them as such, despite the fact that slavery was long gone.[5] The communities they formed around 1888 and in its aftermath can be seen on map 2.

The cacao plantings around the maroons' houses noticed by Otille Coudreau in 1899 indicate that they had diversified their economy in this period. Manioc remained the cornerstone of their diet, but other crops such as cacao, corn, beans, and coffee and fruits like bananas and oranges were now grown as commercial crops, with the threat of reenslavement gone. Former maroon Benedicto Pinheiro sold his land plot to Brazil nut merchant Manoel Costa in 1910. By then he and his family had probably occupied it for a few decades. "Manioc, corn, rice, beans, sugarcane, small-scale livestock industry and Brazil nuts" were Pinheiro's source of income. The sale deed also detailed the "improvements made in the property . . . : a thatched roof hut . . . , some fruit trees, corrals, and paths into the forest." The collection and commercialization of forest items, hunting, and fishing continued to complement the manioc-based subsistence agriculture practiced on the mocambeiros' farms.[6]

Family bonds helped the maroon descendants to maintain community ties across the different *sítios* and villages, and the frequent festivals of the agricultural and the religious calendar provided opportunities for socialization. In the community of Jamary (Erepecurú River), for example, the end of the manioc planting period (April–September) coincided with Saint Anthony's Day (September 24), when a big festival was celebrated that attracted residents from many other communities. At Lake Tapagem, Saint Sebastian was also celebrated with a festival on January 19 and 20, marking the beginning of the Brazil nut season. On January 2, 1931, a resident of the Terra Preta hamlet (Cuminá River) named Olympio Magno da Silveira was celebrating "a *ladainha* or religious festivity followed by a dance," when a brawl broke out. In the investigation that followed, ten individuals who had attended the festival explained to the judge that they came from places as distant as the Acapuzinho and Trombetas Rivers, Erepecú Lake, and the Serrinha village, near Terra Preta, where da Silveira resided.[7]

Some aspects of these religious festivals, such as the singing of prayers in honor of the saints, were shared among the broader rural population, generally known as caboclos, but others were unique to the *mocambeiros*. The periodic celebration of raucous parties at the Stone Shed, with its symbolic importance for the Trombetas populations; the use of the *ramada*, a hut without walls similar to the houses where *candomblé* ceremonies took place in other parts of the country; and the emphasis on playing drums and clapping in these festivities, all had clear African and Afro-Brazilian roots. The songs and the dances themselves often referred to the narratives of origin of the maroon descendants. In the Curuá River area, for example, the Pacoval mocambeiros also held every year the festival of Marambiré, a King of Congo dance celebrating their shared African roots.[8]

By the late nineteenth and early twentieth centuries, some non-mocambeiro colonists had established themselves along the shores of the Lower Amazon rivers; to prevent them from taking over the most fertile land, several maroon descendants began traveling to the parish of Oriximiná to register their homesteads though the so-called *títulos de posse*.[9] The *títulos de posse* or possession deeds were claims recorded in a public registry (local parishes for most of the nineteenth century) by those who had been occupying land plots customarily. According to the 1850 Land Law and the Republican laws of 1890–1891 the *posse* titles did not grant full legal rights of ownership, but were a first step toward achieving them. However, in practice, whether used for commercial transactions or to support claims to ownership over the course of legal disputes, posse deeds were often seen by the mocambeiros as a surrogate for property titles.[10]

The posse deeds also document the mocambeiros' migrations across the sítios and communities they established in those years. Francisca Maria do Carmo was born in the Turuna mocambo in the mid-1800s. She moved to the Colônia settlement and, after marrying Rafael Printz do Carmo, relocated again to the Arrozal Lake, where she registered a posse called Bom Jesus on August 4, 1899. Rafael and Francisca's daughter Sebastiana married José Viana. His father Miguel Viana, also a former maroon, registered a posse called Nazareth on the eastern shore of the Trombetas River on the same day. Florêncio Antonio dos Santos was born in the Campiche mocambo at some point in the 1870s and registered a posse called Santa Maria on the Jacaré Lake in 1898. In 1907, he sold it to Brazil nut merchant Raimundo da Costa Lima and moved to Tapagem Lake, where as an old man he was interviewed by the German missionary and anthropologist Frei Protásio Frikel decades later.[11]

In the meantime, the growth of the regional economy in those decades stimulated the expansion of livestock ranching in municipalities such as Óbidos, Alenquer, and Monte Alegre. Traditionally located closer to the Amazon's main course, a handful of large cattle fazendas also existed at Lakes Salgado, Acapuzinho, and Aracuã up the mid-Trombetas and the mid-Cuminá areas by the 1900s.[12] Cattle ranchers also started to be elected as councilmen in cities and towns across the region; in that role they directed municipal governments to fund expeditions in search of the *campos gerais*, the large grasslands that supposedly existed beyond the waterfalls of the upper course of the rivers. No fewer than six such expeditions, most using mocambeiros as guides and specialists in traversing the waterfalls, were sent to find those grasslands between the 1860s and the 1920s.[13]

Changes also occurred at the state level. As rubber barons filled the seats of the state legislature, the assembly removed the legal obstacles to the extraction and commercialization of multiple forest products, not only rubber. Thus, in 1909 Pará authorized the sale of public land containing Brazil nut trees through Law 1108, which abolished the long-standing tradition of considering Brazil nut groves a public good or *bem de serventia pública*. Several *castanhais*, nut fields or groves, had already passed into private hands through private transactions that described them as being used for an activity other than the cultivation of Brazil nuts.[14] However, this legal change undoubtedly paved the way for the ensuing economic ones.

The colonization of the Lower Amazon river area gained momentum in 1913, when the massive arrival of British latex in international markets led to the collapse of rubber production in Amazonia. Grown in plantations and not extracted in the wild, British rubber from Malaya was cheaper to produce than its Amazon counterpart, and therefore more competitive.[15] After the rubber crash contingents of workers lost their jobs and started to roam Amazonian cities. As the city council of Alenquer explained in a report to the state government in 1921, unemployed rubber tappers arrived in that municipality "from other Paraense counties . . . in the most extreme poverty" and "in search of a remedy for their atrocious suffering." The remedy was to be found in "the abundance" of Brazil nuts, argued the councilmen.[16] For the labor-hungry Lower Amazon commercial houses, which had the infrastructure, capital, and expertise to extract the Brazil nuts, the wandering ex-rubber tappers were nothing but idle hands ready to go collect the nuts.

Although Brazil nuts had been consumed in the Lower Amazon since the colonial period, their price and the volume of their exports steadily increased in the late nineteenth century and then boomed in the early twentieth. The price per hectoliter increased to nearly 20,000 réis between 1890 and 1910 and then skyrocketed to almost 90,000 in 1925; the high prices were maintained until the end of World War II.[17] Coupled with the collapse of rubber, this represented an intense stimulus to the nut exports, as shown in graph 5. In 1935, a book published in Belém remarked how "the shelling of Brazil nuts, and their export, principally to the United States of America, was initiated a few years ago, and is now well established," as attested by the more than 660 tons sent that year to the United States alone. Other European countries started to import the fruit in large amounts by the 1920s as well.[18]

The nuts themselves are the edible seeds of *Bertholletia excelsa*, a giant tree of the Amazon forest that grows in unflooded land and reaches a height of nearly fifty meters and an average diameter of 1.3 meters. Its "fruit consists of a round pod with a hard shell one-quarter of an inch thick, and about four to five inches in diameter. It contains from sixteen to twenty seeds." In the Lower Amazon, the collection of the pods started in December, when the rainy season approached its end and rainwater had made the pods heavy enough to fall to the ground. The workers gathered them at a daily rate of approximately 700 to 800 pods, which represented about ten hectoliters of nuts. The extractors would then open the pods with a machete, extract the nuts, and either peel their hard cover or leave them as they were. At that point they would be either consumed or delivered to a merchant by boat or by mule.[19]

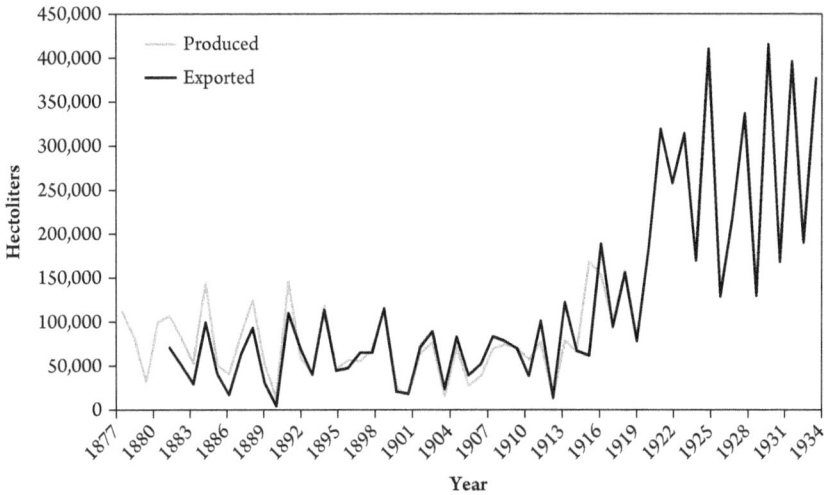

GRAPH 5 Brazil nut production and exports from Pará, 1877–1935 (hectoliters).
Sources: Frazão, *Castanha do Brasil*, fig. Estatistica comparativa das safras, Praça de
Belém; Le Cointe, *L'Amazonie Bresilienne*, 2:460–63; Pará, *Mensagem . . . 1924*, 30;
Pará, *Mensagem . . . 1925*, 128.

One characteristic of the Bertholletia tree is of special interest to our story:
its "highly clumped spatial distribution." The tree is found almost exclusively
in castanhais, groves of 75 to 150 or more trees with a density of 1.5–5.6 trees
per hectare; occasionally the groves were even more dense. Evidence from ge-
netic, linguistic, and archaeological analysis suggests that this pattern of dispersal was at least partially due to human action during the last millennia. The
ancestors of the present-day Arawak, Carib, Tupí and other ethnic families
would have carried the Bertholletias from northern Amazonia to its southern
and southwestern regions. Rodents, monkeys, and other mammals spread the
seeds, increasing the density of Bertholletias in those new areas.[20] It is actually possible to plant Bertholletias and have them produce nuts within less than
a decade, but a widespread belief in the early twentieth century seems to have
been that they became productive only after fifteen to twenty years—and therefore that productive groves would take too long to grow.[21]

This clumped distribution explains the rush to demarcate the Trombetas
lands in the 1920s: once prices for the nuts soared, it became urgent to control
the main clumps of Bertholletias. Data from a collection of 159 official land
deeds (*títulos definitivos*) issued between 1877 and 1940, including deeds
based on ancient *sesmarias* or royal land grants from the colonial era, indicate that thirty-nine, or 24.5 percent of them, were issued before 1900.

Another 18.2 percent, or twenty-nine properties, were purchased between 1900 and 1919. But 43.4 percent of all titles (69 of 159) were purchased between 1920 and 1930, when the price of the nuts and the occupation of the Trombetas by the commercial houses reached its peak.[22] While the demarcation of nut groves had increased in frequency by the 1910s, the height of the nut rush took place between 1920 and 1930.

Going Nuts over Nuts: Land and Labor in the Castanhais

The strategies used by Raimundo da Costa Lima, one of the most active Brazil nut exporters in the Trombetas River area between 1900 and 1930, illustrate how they sought to control the extraction and export of this item. Costa Lima purchased the castanhal São Manoel at the Erepecú Lake, most likely from a former maroon, on July 3, 1899. Before the sale the mocambeiros could collect nuts from that castanhal and sell them freely; afterward they could only collect the nuts if they sold them to Costa Lima at the price he established in the commercial store, the *barracão de comércio*, he built nearby.

To supplement his income and obtain firsthand information about local castanhais, Costa Lima also worked between 1900 and at least 1913 as a land surveyor demarcating a number of nearby properties.[23] Based on this knowledge, Costa Lima purchased a number of posses registered by mocambeiros. In 1906, he bought the Sucuriju posse from Jesuino Helisberto dos Santos and, in 1907, Antonio Domingos de Araujo's Jacaré property.[24] Although in the written record these transactions look legal and fair, in the oral record they were described as fraudulent: maroon descendants claim that these posses changed hands as repayment for a debt or were sold at a much cheaper price than they were worth. The Sucuriju property, for example, was purchased "for 200 réis," the value of a debt previously held with Costa Lima.[25]

But how did these supposedly illegal land acquisitions work? In theory, the legal process governing acquiring a *título definitivo* allowed those affected by the transaction to protest, which could lead to a review or even to the annulment of the process. Application for an official land deed started with a letter to the State Land Bureau (Repartição de Obras Públicas, Terras e Viação) stating that the applicant had a provisional land deed, sometimes simply a posse deed, and wished to obtain a definitive one. The local office of the State Treasury (Mesa de Rendas do Estado) had to certify that nobody else was cultivating or residing on the requested plot. The Land Bureau then announced the date when a land surveyor would demarcate the land by posting a copy at the local office of the State Treasury for thirty days and by publishing it in the state government newspaper (*Diário Official do Estado*). Extra copies of the announcement were sent to the owners of adjoining properties. If there were no objections during those thirty days, then a representative of the Treasury, a land surveyor, a notary, and the applicant met on the designated day and place to start the measurement and demarcation of the property. They read the application aloud, and if no protests were made, they proceeded to survey it, placing markers on its boundaries. Once finished, the land surveyor sent the entire file to the Land Bureau, where an official reviewed the process and charged the applicant the fees for the publication of the announcement, the survey, and the price of the land.[26]

Costa Lima and other merchants knew very well the weak spots in this process and sought to manipulate them to their own advantage. They could, for example, declare in their application that the purchased area was smaller than it actually was. However, surveyors and state officials apparently did not let go these infractions unpunished: merchants were often required to pay for undeclared land and fined for their apparent deception. For example, in 1917 Costa Lima applied for a title to land near the Jacaré Lake, and after it was granted, he had to pay 165 milréis for additional undeclared acreage. In 1924, he had to pay more than four contos for the same reason when demarcating the Cuicé grove.[27] At the survey of the Paraíso property in 1921, he had argued that he "did not know the extension of the property with precision, to the extent that he used the expression 'more or less' when he filled out the application." Other merchants paid similar fines.[28]

Knowing that treasury officials would not always visit castanhais that were located days upriver to check the accuracy of the applications, the merchants also denied that mocambeiros lived on their properties, claiming that their huts belonged to seasonal workers. By doing so, the merchants sought to avoid potential lawsuits brought by long-time residents. When in 1925 Manoel Costa acquired the Uxizal property, located on the Acapú River, he claimed that no workers lived there, but resided instead in São Benedicto. However, when he purchased the São Benedicto property, he declared that no workers resided there either.[29] In 1917, while applying for a property called Leonardinho, Costa Lima stated that the houses of some members of the Cordeiro family and other maroon descendants were simply "huts" (*abarracamentos*) used by nut workers. In 1924, when the lawyer Abel Chermont requested to demarcate a plot of public lands that included a castanhal used for decades by mocambeiro Sebastião Cordeiro da Silva, he followed a similar strategy.[30]

Other irregularities in the process of measurement came to public view when disputes between different merchants erupted. In 1923, for example,

Costa Lima requested that the director of the Land Bureau correct the bound-
ary lines of Extrema, a property that José Gabriel Guerreiro wished to purchase, claiming that "between my properties and those requested by him . . .
there is a distance of thirty or forty kilometers, so the applicant José Gabriel
Guerreiro cannot argue that they are adjacent to each other," as he had done
in his application. The following year lawyer Chermont opposed Costa Lima's
demarcation of the Leonardinho property. After presenting its proven dimensions, Chermont stated that his measurements were taken "with precision,
and without the 'more or less' that is so much to the taste of the applicant Raymundo da Costa Lima," implying that the latter was known for his fraudulent
tactics.[31] That same year Obidense lawyer Elysio Pessoa de Carvalho sent a
letter to the Land Bureau claiming that the demarcation of the Tres Barracas
property had been fraudulent and accusing Costa Lima of multiple irregularities. He wrote that the surveyor Archimino Pereira Lima "is a spiritist, and has
accomplished more than the spirits," by having demarcated a property without visiting it. The application announcement "has not been posted for thirty
days, but for five," and the property was determined to have a perimeter of
54.8 km after a demarcation process that only lasted two hours. In Alenquer,
two 1910 deed applications requested plots that each were claimed to be 3 km
in length, while in reality they measured 22 and 12.5 km, respectively. Due at
least partially to blatant fraud, the result of the competition for natural resources
was most often "the victory of the merchants," as Otille Coudreau sharply
observed about a 1901 conflict over some seringueiras.[32]

The privatization of the Brazil nut groves struck directly at the mocambeiros' principal source of cash income, forcing most of them to work for the
new landowners. Other extractive products were available, but none had the
value of or were as abundant as Brazil nuts. By early January, at the beginning
of the collection season, most men, some women, and even entire families
moved to the castanhais to work.[33] At Lake Erepecú, where some of the largest castanhais were located, groups numbering several dozen individuals
relocated together for the harvest season.[34] One maroon descendant recalled,
"When the time came, all of them [the mocambeiros] left," often for the nearest castanhal "to work." Those living in the Tapagem, Boa Vista, and Jacaré
communities would go to the Erepecú Lake area or to castanhais like Fartura
and Paraíso, belonging to Costa Lima, while those living at Mãe Cué and nearby
areas would go to Cuicé.[35]

When the mocambeiros arrived either alone or with their families to the
castanhal of a known *patrão*, they requested a *colocação* or work assignment.
They set themselves up in a hut provided by the patrão, "gathered [the nuts],

FIGURE 4 A mocambeiro Brazil nut collector in the Curuá River in 1900. He is carrying a big basket strapped to his head and back to collect the pods. Coudreau, *Voyage au Rio Curuá*, 53. Courtesy of University of Wisconsin-Milwaukee Libraries.

peeled them, and brought them to the *paiol*," a shed where the nuts were housed. Nearly every Saturday they brought the weekly production by boat or by mule to the owner's shed—usually located strategically to control the waterborne traffic—where it was weighed and recorded in the *livros de aviamento* or accounting books. The workers also purchased the staples they needed, mostly manioc flour, from the commercial store or *barracão de comércio*, as well as varied goods and foodstuffs like cloth, hammocks, steel tools, fishing equipment, guns, gunpowder, salt, coffee, sugar, poultry, meat, and soap. Normally, the farther their permanent home was from the castanhal, the more goods they would have to purchase. A company owner or a supervisor oversaw the shipping of the nuts to the city and monitored the river traffic from the barracão, often located on the dock from which nuts were shipped to the city. The main threat was the presence of competing merchants navigating the rivers and streams close to the castanhais, as that often meant that the extractors were practicing contraband.

Once a month or at the end of the season, the merchant calculated the *saldo* or balance of the goods purchased versus the amount of nuts each extractor brought to the barracão. If the worker earned more than he purchased, he was paid in cash.

If at the end of the season the balance was negative, he had to work the next year for the same patrão, or could repay him by gathering other products during the year. As in the case of rubber, this system was called *aviamento*: the worker was the *aviado* or *freguês*, and the supplier of the goods the *aviador* or patrão.[36]

This basic design, however, admitted numerous variations. Some merchants operated from the central offices and stores of commercial firms in Oriximiná, Óbidos, or Alenquer, supplying goods to smaller independent aviadores in exchange for their fregueses's nuts or even buying nuts from independent, individual extractors. There could be as many intermediaries as the price of the nuts allowed. At the upper end of the chain, the aviador would sell the nuts to an export house that shipped them to foreign markets, mainly the United States and Europe.[37]

As the trade expanded in the 1920s and 1930s, new castanhais were demarcated farther north, in the Acapú and the Upper Cuminá regions, and it became necessary to bring workers from nearby populations and municipalities. As Zé Melo from Boa Vista explained, "The locals were not enough for the harvest. . . . So in those castanhais upriver . . . they brought people from Oriximiná, from the Apocú . . . from Terra Santa, from Faro." Luis Guerreiro's grandfather José Gabriel Guerreiro brought workers from his ranches downriver. During the dry season they worked as cowboys and agricultural workers, but between January and May they were employed as extractors in the castanhais of the Acapú River area. In 1926, Elyzio Pessoa de Carvalho requested the seizure of Costa Lima's nuts extracted from the Ultimo Ponto castanhal on the Upper Craval. During the ensuing trial over the property of the nuts, only one of the six workers who testified was a mocambeiro—the other five were extractors brought in from Oriximiná.[38]

Maroon descendants claim that the relations between merchants and collectors greatly favored the former. The price of food and implements was very high, producing negative balances for the castanheiros. According to Zé Melo, who started working as a castanheiro in the mid-1950s at Lake Erepecú,

> We knew that they were exploiting us, but we had no choice. The manioc flour . . . many people did not bring theirs to the castanhais, and then you had to purchase it there. Sugar and everything else was more expensive there. . . . At the end [of the nut season] you wanted to pay the balance of the products you bought, [and] the produce you had collected, the nuts, was not enough to cover it. You were in debt. . . . The nuts were cheap, and the products were expensive. You worked just for food in those months, most times you did not manage to bring home anything.

Dona Biquinha, who worked at the Cuminá castanhais in the late 1940s, explains that "it was hard" to end the season with cash, "because it [the Brazil nut] had a low price. . . . You collected many nuts but made little money." When Maria Teresa Fernandes Regis's father died in 1947, she started working in the castanhais of Lake Erepecú, and like Maria Rosa Xavier Cordeiro, from the Tapagem community, who became a castanheira in the late 1930s, she recalls that the nuts fetched a low price at the commercial store.[39] Valério and Zuleide dos Santos, from the Boa Vista community, who worked with their families for Costa Lima and his descendants at Jacaré Lake, claimed that he "cheated people. He never calculated the balance in front of the employee; when the worker got there, the balance was ready." Added Zé Melo, "Yes, that happened. Because look: we did not know mathematics, and they cheated people. In addition to buying the produce cheap, they did the math and kept everything."[40]

While it is difficult to assess the extent of this cheating by the merchants, because it did not leave any documentary evidence other than the periodical conflicts it caused, there is evidence that they resorted to coercion in order to enforce their control of this system. Former castanheiros, for example, contended that if the owner of a castanhal or his manager caught a gatherer trying to secretly sell nuts to a different merchant or to a regatão, the worker was either arrested by the police or beaten by the manager. "You went straight to jail," explained Maria de Souza, from the Erepecurú. "If you sold [nuts] to a regatão, you went to jail," remarked Maria Rosa Xavier Cordeiro, to which her brother Seu Duí added that "you lost the nuts, and you lost the job. They never gave it back to you."[41]

The police acted swiftly to defend the merchants' interests when they reported thefts of nuts. Antônio Souza, a Brazil nut worker during the 1950s and a mocambeiro descendant from Lake Abuí, claimed that when a patrão accused a local of stealing nuts, "before a week had passed, the police came, caught them, and sent them to Oriximiná." Exactly seven days is indeed what took the Alenquer police chief to arrest and seize the nuts collected by "strangers invading the castanhal [Felinto], where they are extracting nuts without the authorization" of the owners in January 1936. On March 6, 1930, Portuguese merchant Joaquim Tavares de Souza, who controlled the Brazil nut groves adjacent to the Pacoval community in the Curuá River area, complained to the police that one Francisco Paes had "invaded and gathered a large number of nuts" from his castanhais. "You should immediately summon the invader and instruct him to cease these actions," the Alenquer police chief then instructed the local officer. "Otherwise, he will be brought to this office" to be interrogated.[42]

Sometimes the merchants' tight grip over the extraction and selling of the nuts expressed itself through physical violence. Manoel Vicente de Oliveira, a worker who had recently arrived at the castanhais of Alenquer, was "violently incarcerated" when a merchant reported to the police that he was working in a nearby city to repay a debt. In July 1935, Alenquer police chief Columbiano Marvão relieved Antônio Vianna dos Santos, the manager of a Brazil nut commercial house, from his duties as police deputy in the maroon descendant community of Pacoval for "exceeding his duty." Dos Santos had beaten Antônio Rufino de Oliveira, a local nut worker who had allegedly stolen nuts from him. He was suspended until an investigation was concluded; the results of the investigation remain unknown. In sum, the Brazil nut merchants successfully enlisted the might of the law to their cause and, at least occasionally, resorted to acts of "violence and outrage" as well.[43]

Taking Revenge: Turning Nature into an Ally

For the mocambeiros, it was not easy to cope with the privatization of the Brazil nut fields. Since the colonial period the availability of marketable forest goods and of land to cultivate manioc and other foodstuffs had remained at the heart of the local peasant way of life, embraced by maroons and rural slaves alike, because it gave them the possibility of forming autonomous family groups and communities. Forest goods were not only a part of this equation—they were key to it, because they represented access to cash and therefore were a fundamental complement to the cultivation of staples. Rural Amazonians perceived extractive products as a public good whose usufruct was regulated by the community and the household, in accordance with indigenous and caboclo social traditions.[44] Therefore, the mocambeiros perceived outsiders' efforts to establish new rules to access forest and river products as illegitimate, unfair, and abusive, and they employed an array of strategies to circumvent, undermine, or confront the merchants' takeover of the nut fields.[45]

While migrating to nearby cities like Oriximiná, Óbidos, Alenquer, or Santarém may have been one solution, the mocambeiros, even the young ones, did not do so very frequently between the 1920s and the early 1960s. "People at that time did not like their children to live far away; they wanted to keep them close," explains one maroon descendant. Mistrust of rich landowners may have motivated this refusal to permit offspring to leave: the locals knew that their children could end up under the legal tutelage of richer individuals, a deeply ingrained fear among those who had known slavery just a few decades ago.[46] It was not until the later 1960s and 1970s that migration fluxes to Lower Amazon cities became stable, as the labor market expanded and access to health care and education for Brazilians increased. But until then, rural to urban migration in the area was apparently not very attractive.

Instead many mocambeiros tried to locate new Brazil nut fields that they could keep free of the merchants' control. Seu Duí related how in 1949 "my father had a *ponta de castanha* [small grove] that he called Encantado. It is located outside [Zé] Machado's [Costa Lima's son-in-law] Brazil nut grove. Only he and other elders knew about it. It was far away from his workpost." When Machado reported to the police that Seu Duí and his father had stolen nuts from the merchant's property, the Cordeiros indignantly rejected the accusation. "Negative, young man," the father replied to the policeman sent to seize the nuts. "I will not give you the nuts. They are mine, we got them from that ponta de castanha where we work; it does not belong to Zé Machado." After a long discussion, the policeman finally carried the nuts away under the promise that Machado would compensate the Cordeiros for the seizure. Ultimately, however, Machado "did not pay for the nuts. He kept them, but did not pay. They were gone," Seu Duí remarked with sadness. His case was not unique: other maroon descendants tell similar stories of merchants seizing nuts collected from castanhais that did not belong to them.[47]

Another possibility was to "take revenge" on the merchants by pilfering nuts and selling them to other buyers, be they regatões or competing merchants. This practice was known among former castanheiros as *driblar*, to dodge or to get past the merchant's watch. It was apparently widespread among the maroon descendants, because they considered that "everybody had the right to extract nuts to sell them," as Pacoval community's Dona Piquixita explained when discussing the causes of a conflict over nuts in the mid-twentieth century. Others related how many individuals "extracted nuts behind the merchants' back," often at night, and "sold them to another [regatão or merchant] who paid a better price."[48] Captain Manoel Campello de Miranda explained to the state chief of police in September 1931 how "I am constantly asked to take measures against the abusive habit of many individuals who are hired to work in the Brazil nut harvest . . . and then they sell [the nuts] to others furtively." In response to these acts of pilfering, the seizure of nuts by the police increased during the 1920s and 1930s.[49]

The growing competition among commercial houses led some commercial agents to engage in dishonest strategies against their rivals. Seu Duí, for example, recalls how in the late 1940s Italian merchant Braz Sarubbi had an

employee who bought nuts from Zé Machado's workers at the Farias castanhal near Lake Tapagem: "he entered at night and bought the nuts," thus
putting profit before the law.[50] It was easier to make such illegal purchases
in castanhais closer to populated areas, where the denser river traffic and the
proximity to trade houses provided more opportunities for regatões to buy
nuts at night or in clandestine transactions. In castanhais near the maroon
descendant community of Pacoval, for example, pilfering was more frequent
than in the remote Acapú area, where the traffic of people and canoes was less
dense and therefore could be controlled more easily.

In sum, as the merchants employed higher degrees of coercion, the
maroon descendants responded with a series of strategies of resistance that
replicated past relationships with the river's natural landscapes. Facing coercive and occasionally violent forms of labor control, the mocambeiros used
the region's intricate watercourses to find hidden nut groves and sell nuts
behind the back of the commercial managers or *fiscais*. Such actions were
not so different from those employed at the time of captivity to flee the slaveowners: in both cases the natural environment was instrumental to evad-
ing social control. It is little wonder that the new reliance on nature gradually
merged with the old one in the social memories of the mocambeiros, giving
substance to the idea of the "second slavery." As it often happens in the oral
histories of rural African descendants all over Brazil, memories about the
post-emancipation decades blended with those from the time of slavery,
thus forming a single experiential compact centered on exploitation and
poverty.[51]

Becoming Bootlickers: Compadrio and Accommodation

This increased commercial competition also led some merchants to develop
relations of patronage with certain individuals and communities. As shown by
the cases described in chapter 2 of maroons who traded with unscrupulous
merchants "in broad daylight" during the last years of slavery, trade was already
at the heart of relations of patronage between maroons and members of the
local elite. The maroon descendants of Pacoval, in the Curuá River region, used
Brazil nuts to develop patron-client ties with some members of Alenquer's economic elite even before abolition, as did the Trombetas maroons who traded
nuts, tobacco, and other items with commercial houses in Óbidos in the 1870s,
and perhaps even earlier.[52]

The first individuals to engage in alliances that crossed the boundaries of
class and ethnicity were the specialists in finding undiscovered castanhais in

the middle of the forest, known as *mateiros*. Costa Lima, for example, hired the maroon descendant mateiro Augusto Cordeiro to help him discover new Brazil nut fields in the early twentieth century. A grandson of runaway slave Caetano Antonio do Carmo, Cordeiro was born in 1896 near Lake Abuí.[53] Present-day maroon descendant Manoel Francisco Cordeiro Xavier, aka Seu Duí, explains that "there was this person in my mother's family called [Augusto] Cordeiro, who was a great explorer for Costa Lima. So Costa Lima made the preparations to go into the forest looking for Brazil nut fields . . . they walked for days, found a ponta de castanhal and [Cordeiro] gave it to him [Costa Lima]. Since he was an engineer [i.e. land surveyor], Costa Lima went there to demarcate the castanhal. And from then on it belonged to him."[54]

Several Trombetas maroon descendants also recounted Augusto Cordeiro's role as an explorer and discoverer of castanhais.[55] Even Luis Bacellar Guerreiro, the son of prominent Brazil nut merchant José Gabriel Guerreiro Júnior, identified Cordeiro as "Costa Lima's explorer." Other merchants benefited from the knowledge of mateiros like Raymundo Serrão at Lake Erepecú, Manoel do Nascimento in the area of Boa Vista, and João Petronillo Pereira, who discovered the Cruzeiro, São Sebastião, Bella Brisa, Boa Esperança, and Tres de Abril castanhais in the Craval River for merchant Elysio Pessoa de Carvalho in the late 1910s and early 1920s.[56]

Another special relation was created when maroon descendants became the merchants' commercial representatives and *prepostos*. A preposto was typically an extractor or an administrator who settled in an unoccupied area on behalf of a Brazil nut trader; the presence of a preposto legitimated a merchant's claim to buy and demarcate that land. Most contracts between prepostos and their superiors were informal, and to the best of my knowledge none has survived in the archives of the region. The services of prepostos were frequently used in disputed areas containing nuts: in 1926, for example, Elyzio Pessoa claimed that Costa Lima had sent prepostos to occupy a castanhal *near* the Upper Craval River, although Pessoa himself had employed the Magalhães brothers as prepostos just a few years before.[57] Sometimes the job of preposto merged with other professional categories in both the written and the oral record, creating a hodgepodge of labels defining favored employees, such as *fiscal* (manager), *mandatário* (envoy), *representante* (representative), *procurador* (attorney, solicitor), *explorador* (explorer), and even *freguês* (customer). Maroon descendant Manoel Theodoro dos Santos, a former preposto who appears as "owner of a commercial house" in tax records from 1907 and 1916, was actually Costa Lima's employee or fiscal in different castanhais, although he was also described as his freguês in other documents.[58]

Finally, some merchants sought to gain prestige among the mocambeiros by becoming their godfathers. As Charles Wagley discussed in his 1948 ethnography of social and cultural life in the Amazon town of Gurupá, when parents choose a godfather and godmother to sponsor their child's baptism, the godparents enter a web of tangled relations "of mutual respect, of mutual aid, and of intimate friendship" with their godchildren. Poor rural dwellers often asked "a trader, a public official, or anyone with prestige and a relatively superior economic situation" to become godparents of their children. While the parents hoped that "the co-father of higher status and greater wealth will favor and protect his poorer co-father," from the godfathers' perspective "the number of godchildren and co-fathers which an individual can claim is an index of social position."[59]

Seu Duí, whose godfather was Zé Machado, Costa Lima's son-in-law, explained how when he was born "they [Zé Machado and his family] asked to baptize me there, at Lake Jacaré, and so they did." Seu Duí's sister Dona Maria Rosa had Judith da Costa Lima, Raimundo's daughter, as a godmother. Merchant Milton Magalhães had a store near the community of Pacoval. In 1921, he sided with the Pacovalenses in a public protest during the visit of a state commission to resolve a land dispute. His relations with the mocambeiros were not simply commercial. He was also the godfather and wedding witness of Dona Piquixita, a maroon descendant woman from Pacoval. Bonds like these continued a long-standing local tradition between maroon descendants and the nut traders.[60]

There were, for sure, some rewards to be gained from these relations of favor. An illustrative case is that of Antônio Vianna dos Santos, from Pacoval (municipality of Alenquer), who by the 1920s was Portuguese merchant Jaquim Tavares de Souza's main manager in the Felinto and Praia Grande Brazil nut groves.[61] He also became de Souza's *compadre* (godfather). Vianna lived near the groves and closely supervised the maroon descendants who gathered the nuts. Sometimes he was too zealous in carrying out his duties: as indicated earlier, he beat at least one, and probably more, of those who attempted to pilfer nuts. From Souza's standpoint, Vianna was a very valuable asset: because he was a local, he had privileged access to information about attempts to defraud the owners of the castanhais. Vianna thus was able to earn money without having to work as a castanheiro and was named the godfather of numerous children in the mid-1930s, an indicator of considerable social prestige among his fellow Pacovalenses. But his high status yielded even more benefits: when his granddaughter Figênia became pregnant by one of Souza's relatives in the early 1960s, he was compensated with ten head of cattle thanks to his special relation of compadrio with Souza.[62] Later, the illegitimate father provided funds for the education of that child.

However, as can be inferred from Seu Duí's description of the relations between Costa Lima and his mateiro Augusto Cordeiro, in the end the relationships between merchants and their favorites were essentially unequal. Costa Lima "paid a reward" for the explorer's service, although eventually he "became the owner" of the land he discovered. Other maroon descendants confirmed the unequal nature of this relationship. According to Deometilo Cordeiro, Augusto Cordeiro's grandson, the merchant "asked the mateiro to look for the land [containing nuts], keeping all the land for himself. The explorer arrived and said, 'Look, I found a nice Brazil nut field with these trees, it could yield these many crates.' And the merchant responded, 'Where is it? Let's go there to demarcate the terrain.' And they went there to demarcate the land for the merchant. So the explorer kept losing the lands, and when he looked for some, none was available." A slightly different version, although with a similar outcome, was offered by Ruy Brasil when discussing Costa Lima's strategies: "when a black discovered a Brazil nut field, old [Augusto] Cordeiro had to follow him to see the location. When Cordeiro discovered it, he came back and told his compadre Costa Lima." Since he was a land surveyor, the land was rapidly demarcated, and whoever took Costa Lima there was told "Hey, if you sell these nuts [to somebody else], you will go to jail, because this *castanhal* is mine."[63]

Fictive kinship relations such as that of compadrio between merchants and mocambeiros buffered conflicts between them. In the case discussed earlier, the fact that Portuguese merchant Tavares de Souza could compensate Antônio Vianna with payment in cattle head made possible a negotiated exit to a potentially conflictual situation. Had these ties not existed, the Viannas ran the risk of obtaining no help whatsoever and even of being publicly humiliated, as happened when merchant José Antonio Picanço Diniz impregnated a female worker on his family's ranch in 1934 and then was acquitted at the trial for the recognition of his paternity.[64] Compadrio relations, then, helped ease the strains derived from class, ethnic, and even gender domination, while simultaneously reinforcing the social prestige, the moral authority, and even the material welfare of the maroon descendants who became compadres of their patrons.

In exchange for the loyalty of the mateiros and other favored employees, and to increase their influence as godparents, some merchants also provided money and goods on credit to their employees to enable them to celebrate events as important as baptisms, weddings, saints' days, and the end-of-the-harvest parties in June.[65] By inserting their public displays of patronage onto the special occasions where the mocambeiros reproduced their collective values and practices, the merchants sought to counterbalance their image of exploiters with that of "fathers of the people," as Costa Lima was called by a maroon descendant.[66] In the backlands of Amazonia manufactured items were expensive and hard to obtain, and command over their distribution represented a valuable source of power and influence—one that Costa Lima and others cunningly used to present themselves as benefactors.

Providing funds for celebrations also enabled merchants to appear as patriarchal leaders, not unlike the classical political patrons or *coronéis* of rural Brazil during the periods of the empire (1822–1889) and the First Republic (1889–1930).[67] While rural landowners amassed their wealth by accumulating land, slaves, and control over state institutions, early twentieth-century Amazonian patrões like Costa Lima tried to mimic the patriarchs of rural Brazil by weaving local networks of favor, clienthood, and trade with local populations. The merchants' zeal in protecting their command over the supply of manufactured and locally scarce items, and their establishment of rituals of deference and obedience on the basis of that supply, only make sense if we see them as part of the merchants' attempt at becoming *coronéis*, the classical patriarchs of rural Brazil.[68]

In the end, this network of favor and patronage between mocambeiros and nut merchants served well the interests of the latter. While it allowed some mocambeiros to improve their situation, it also enabled the traders to advance their agenda by creating a body of reliable managers and prepostos who provided firsthand information about new castanhais and any illicit activities engaged in by the workers. The merchants could thus penetrate the cultural fabric underpinning the maroon descendants' expertise in nut extraction and curb their resilience to external control, thus eliding some of the tensions caused by the process of land privatization. In the long term, by providing an "escape hatch" from class, ethnic, and even gender conflicts, and by reducing the mocambeiros' capacity to act independently from the merchants, patronage between Brazil nut merchants and maroon descendants consolidated class and racial domination of the former over the latter.

Conclusion

In the immediate aftermath of abolition, between 1888 and the 1910s, Lower Amazon maroon descendants enjoyed a period of personal and collective autonomy. They were free to settle along the area's rivers, forming villages and farms. Sustained by a peasant economy based on a portfolio of extractive and agricultural activities, the mocambeiros also maintained a vibrant collective culture in celebrations tied to both the religious and the agricultural calendar. While their sítios and their communities had a very similar culture to that of other Amazonian caboclos, the black peasants also displayed pride in their mocambeiro history. Every year they celebrated the Marambiré and other African folkloric events, worshiped syncretic deities at the Big Stone Shed, shared stories about their origins as maroons, and revered community leaders according to cosmogonies of African ancestry. The maroon descendant ethnicity did not fade away with slavery—it flourished with freedom.

However, as the rubber boom came to an end in the 1910s, a number of merchants and landowners moved into the region, acquiring property in an effort to exploit Brazil nuts. As shown by the study of Raimundo da Costa Lima's strategies, the merchants purchased posses that belonged to maroon descendants, sometimes through abusive practices, and relied on their knowledge of administrative procedures to bend the laws regulating land acquisition. They cheated about the size of their properties, made maroon descendants appear as nut workers, and, in general, tried to erase the mocambeiro presence from the documentary record. To secure a stable supply of laborers, the merchants also subjected the mocambeiros to debt peonage and imported workers from outside the area. By fixing high prices at the company store, manipulating the balances, and relying on the help of police forces, the merchants kept their workers under their control and secured a disproportionate share of the profits of the Brazil nut trade.

However, the mocambeiros did not accept this reality passively. They pilfered as many nuts as they could and sold them under the table to the wandering regatões and to competing merchants, using intraelite rivalries to soften the effects of their subjection. They tried to keep some Brazil nut groves secret in order to safeguard some cash-producing resources for themselves. In this sense, early to mid-twentieth-century disputes over command of the landscape reproduced the contours of past conflicts, renewing the value of an advanced knowledge of natural landscapes and reinvigorating the memories of the oppression suffered by the mocambeiros under slavery.

Despite the mocambeiros' shared racial, ethnic, and class profile, it would be an error to conceive of them as a cohesive and homogeneous group. Some carved out their own special relationships with the merchants, thus obtaining extra cash payments, better wages, contributions to community celebrations, better prices for nuts, social prestige, and protection from some abuses. They became in-betweens of the Brazil nut trade, forming a small, albeit significant, category of middlemen who inhabited the liminal space between white merchants and black peasants. For the merchants, the existence of environmental experts who could be turned into local allies in the form of mateiros and other skilled employees was extremely useful, constituting an ideal tool to redraw the impenetrable geographies of black autonomy that had existed to date in the rivers of the Lower Amazon.

Social relations in rural Amazonia during the First Republic were thus not that different from those of other regions in Brazil, where political chiefs combined traditional patronage with a reliance on institutional power to reinforce their influence among rural populations. At the level of accommodation and collaboration between merchants and nut extractors, we find owners of commercial houses who, seeking to smooth over their penetration into local landscapes, tried to appear as benevolent patrons of the community by sponsoring parties, baptizing individuals, and becoming compadres. Indeed, some mocambeiros remember them as "fathers of the people" and benevolent patrons.

To a great extent, the merchants won the game. They amassed substantial fortunes, diversified their activities, and eventually conquered local politics thanks to the wealth brought by Brazil nuts. Take the Guerreiro family, for example. When in 1934 Oriximiná was separated from Óbidos and became an independent municipality, José Gabriel Guerreiro's son Helvécio Imbiriba became its mayor between 1935 and 1943; Helvécio's brother José Gabriel Guerreiro Júnior ruled between 1943 and 1947; and another son, Guilherme, was mayor between 1948 and 1950. They spent a total of fifteen years in power. Raimundo da Costa Lima's son-in-law inherited extensive Brazil nut fields valued at 155 contos de réis in 1941 and also attained considerable influence in Oriximiná.[69]

In the meantime, for maroon descendants the loss of autonomy and access to cash meant their remembering this period as a "second slavery." The alienation from Brazil nuts meant forfeiting a fundamental cash source. While access to the nuts had provided a chance to fight slavery decades before, in 1910 their privatization forestalled maroon descendants from successfully tapping on market resources for the rest of the Old Republic. As of 1942, an appalling majority of black peasants in the Lower Amazon was still illiterate and continued to eke out a modest living from the sale of manioc flour and a handful of forest products.[70]

CHAPTER FIVE

Citizens of Tauapará

Landscape, Law, and Citizenship in the
Senzalas, 1862–c. 1944

In 1944, the rancher Rodolfo Englehard received an unusual delegation on his farm on the island of Marajó. A group of black farmers that had lived for decades on Englehard's Santo Antônio da Campina ranch, in nearby Vigia, came in protest, alleging that a new manager of the ranch had charged rent to the forty-two families living there. That rent was unfair, they argued, because their uninterrupted and peaceful occupation of those lands for a half-century made them legal proprietors: "I was born and raised there, and so were my mother and my father, and all my family," argued Seu Nunhes, one of the tenants. "How come we should pay?" Englehard replied that it was legal to demand rental payments, because the tenants were extracting various natural resources from areas that did not belong to them.

Seu Nunhes then showed him a document from 1908 listing the names of the families living in the community. "No, sir," he insisted. "You either liberate us from any payment or you will pay compensation" to those individuals who decide to relocate. After some discussion, Englehard acceded to their petition: they could reside on that land, cultivate it, and use its natural resources without paying any rent. Most of those families still live on the island of Tauapará, in what is officially known in the present as the *remanescente de quilombo* or black community of Cacau. Accustomed to stories of landless peasants abused by large landowners in Amazonia, it was surprising to hear one describing a victory by a group of black farmers. "Could Englehard have evicted the families from those lands had he tried harder?" I asked. "No," clarified Seu Nunhes, clenching his right fist and waving it assertively, "because all the local authorities knew that we were citizens of Tauapará."[1]

This expression sounded strange, because it mixed a political concept, citizenship, with a specific landscape, the island of Tauapará. I did not understand it initially, but as I spent more time investigating these farmers' history, I came to realize that for Seu Nunhes the concept of "citizens of Tauapará" encapsulated the compact of citizenship rights invoked when the Cacauenses defended their community from encroachments on their lands. If I was to understand how they perceived their rights of citizenship, it was necessary to disentangle the different components of this expression. Taking Seu Nunhes's words as a point of departure, this chapter argues that the black peasants merged slavery-era traditions embedded in the landscape with discourses of legal access to landownership learned in freedom to form a single but multi-faceted discourse of citizenship.

To make that argument, I examine two processes. The first relates to the changes in the landscape that took place as the Campina sugar plantation labor force became a farming community between 1862 and 1944. Landscapes, as discussed in the introduction, are "the symbolic environment[s] created by a human act of conferring meaning on nature and the environment"[2]: in other words, the sum of a physical place plus the narratives people formulate about it. In this chapter I discuss three moments in the history of Campina's landscapes: its "golden age" as a slave plantation, the last years of slavery, and the early twentieth century, as its former slaves completed the transition to free peasantry. The sequence of these three moments illuminates the social and economic changes the slave-descendants experienced, and constitutes the material basis for the second part of the chapter. In that part, I turn to the ideological and cultural origins of the construct of citizens of Tauapará by interrogating popular understandings of landownership, relationships with patriarchal landowners, and the imagined life of legal documents.

The trajectory of the black farmers of Cacau during the decades after emancipation reflects that of other black peasant communities in the rivers near Belém,[3] and explains why currently there are so many remanescentes de quilombo communities along the rivers of the Guajarine basin. After the abolition of slavery in 1888, a number of slave descendants continued to live on the same rural properties where they had toiled as slaves, because most former plantations were in areas that contained no rubber. In sum, the neglect of most of the traditional plantation areas allowed them to become tenants, and in some cases even proprietors, of the lands they had inhabited for generations. It may be more accurate to call these freed people senzala descendants, rather than quilombo descendants, because through a combination of internal and external factors they were able to transform the old landscapes of slave labor into new farming communities.

"Everything about Indicated Opulence and Plenty": Slave Landscapes in Campina, 1846–1862

The Santo Antônio da Campina sugar mill lay on the shores of the Tauapará River, which flows through the island of the same name and leads into the bay of Guajará Miri. Right across from the island, on the opposite shore of the bay, lies the city of Vigia. It is not known with certainty when Campina was built, but most likely it was at the time Pará's plantation sector was taking off in the second half of the eighteenth century. When the Cabanagem revolt shook the region in 1835, the mill was owned by Agostinho José Lopes Godinho, a rich Portuguese merchant who found himself "compelled to flee the country" and so was unable to prevent "the sacking of his place." He ultimately "sustained great loss," explained American entomologist William Edwards in 1846. Like most of its counterparts, however, the plantation survived the Cabanagem revolt, and by the time Edwards visited it, "everything about indicated opulence and plenty," and "two mills constantly employed [in processing the sugarcane] were insufficient to dispose of his yearly crop."[4]

Under Godinho, the Campina plantation included a series of landscapes of slave labor carefully designed to produce not only sugar but also "cashaça [rum], that being considered more profitable than sugar or molasses." The sugar mill relied on the river's tides as its main source of energy, thanks to the almost four meters of difference between the water's ebb and flow in the lower course of the Amazon River. When the tide flowed in, water from the Taua-pará River was stored in a natural pool; when it flowed out, the stored water was directed to an artificial water channel, where it spun the waterwheel that drove the mill rollers. After the cane was crushed, the sugarcane juice was either boiled in the kettle house to make sugar, or stored and distilled into *cachaça* in a large shed—the "two mill houses" of Godinho's 1862 postmortem inventory. The mill's harnessed tidal power was also used to propel watercraft out to surrounding rivers, facilitating the transportation of the sugar and the cane brandy to the city. In 1846, Henry Edwards saw "oxen driving the sugar mills," which probably indicates a seasonal combination of both tidal and animal power to maximize the mill's working time.[5]

The rest of the properties' workplaces were designed around the central position of the sugar mill. The seventy-five slaves working in Campina in 1862 also toiled making bricks, tiles, and clay pots at the *olaria* or brickyard, where "stacks of demijohns and jars were piled in the rooms, or standing ready to receive the cashaça [*sic*] or molasses." Nine cows, twenty-three oxen, four calves, and a donkey grazed in a prairie close to the mill that Edwards found "irresistible." Finally, free workers under tenancy contracts lived with their families in a series of homesteads (sítios) scattered across the periphery of the property, like São Thiago or São José, growing foodstuffs like manioc, corn, beans, and assorted fruits and vegetables. The surrounding forests and mangroves provided an almost unlimited repository of berries, game, and fish—including river crabs, açaí berries, *turú* worms (a delicacy), *pacas*, and agoutis (large rodents)—to supplement the diet of the free and slave inhabitants of Campina. In nineteenth-century Amazonia, large sugar plantations with a diversified production like this one tended to be called *fazendas*, a term that in principle designates livestock farms, not sugar mills.[6]

Close to the sugar mill was the tree trunk where the slaves were flogged, a place that left indelible and traumatic scars in the memory of the locals. Born in 1922 in the sítio of Santo Antônio, Dona Bena visited Campina during her childhood and felt horrified by "the trunk where they punished the slaves." Seu Zacarias, born in 1943 in the same community, relates the stories his grandfather witnessed about the "evil" treatment the slaves received—how the master "put them in the trunk and lacerated their skin" and "when he was done, he threw salt and vinegar" in their wounds. Songs composed in Tauapará during the early twentieth century emphasize how heinous a symbol of oppression Campina's tree trunk (*tronco*) was, describing how "a master had a negro tied up [to the trunk] / with chains in his ankles / the black will be beaten." If the mill's hydraulic works reflected pride in having tamed natural watercourses, the flogging trunk expressed the darkest side of slavery and inscribed it in a specific location.[7]

The *casa grande* or big house, another manifestation of seigneurial power, was located very near to the mill. "Decorated with objects from Europe," it had "two stories and many rooms," as well as "long gloomy corridors, courtyards and windows everywhere" and "heavy doors with iron bolts," explained Dominga de Moraes, a former domestic servant and daughter of household slaves. The second floor was occupied by the planter, his family, and the house slaves; the overseer and his own family probably lived on the ground floor, following a model common in Bahia and Rio de Janeiro.[8] While in the rest of Brazil most masters' houses were located on higher ground, dominating the plantation from a symbolic and material position of power and privilege, this was not possible on the flat terrain of Campina. Even despite its lack of elevation, those who saw the Campina casa grande described it as "very beautiful," "huge," and "seigneurial." The house, made of masonry and boasting a number of luxuries imported from Europe ranging from furniture to gold cutlery, undoubtedly made an impact in rural Pará, where most rural dwellers lived in humble palm-thatched, wattle-and-daub cabins.[9]

Adjacent to the masters' house was a chapel consecrated to Santo Antônio containing an image of the saint brought from Portugal—perhaps from Godinho's own parish of São Martinho da Vila in the province of Alcobaça. The master worshiped a saint venerated by the white upper classes throughout

Amazonia and was a devout member of the Brotherhoods of Nossa Senhora de Nazareth and of Senhor Bom Jesus dos Passos. He also performed a series of religious rites of patronage, thereby using Catholicism as a mechanism of social control: "soon after sunset all the house servants and the children of the estate came in form to ask the Senhor's blessing, which was bestowed by the motion of the cross, and some little phrase, as 'adeos.'" The blessing was meant to express the reciprocal ties between the patriarchal master and his slaves, a statement of hierarchy reminding the slaves who was in command. Both the conferral of the blessing and the use of plantation chapels were common throughout slaveholding Amazonia.[10]

Evolving Landscapes, 1874–1888

On June 6, 1874, imperial congressman Domingos Antônio Raiol purchased the Santo Antônio da Campina plantation.[11] At that time Campina had "an oxen-powered sugarmill . . . houses in a bad state, cattle, tools, [fifty-seven] slaves, improvements [*bemfeitorias*]," and the homesteads or sítios Cacao, São José, Pedreira, São Thiago, and Cumihy, all located very near the *engenho* (mill).[12] The slave crew had decreased in size from seventy-five to fifty-eight individuals; most importantly, by the time Raiol purchased Campina its slaves had developed a series of internal kinship ties that were more characteristic of a community of free peasant families living together than of a plantation slave crew.[13] It became clear that the property's slave crew had begun to alter Campina's landscapes, transforming spaces of work and residence, as well as the slaves' relationships with the natural world.

Whatever obstacles a large African slave population may have represented to the formation of families in the past, they disappeared shortly after the cessation of the transatlantic slave trade to Pará in 1841, making possible a vigorous construction of kinship ties. While in 1862 only ten of Campina's seventy-five slaves, or 13.3 percent, were registered as having kinship ties with one another, by 1874 this number had increased to thirty-eight out of a smaller population of fifty-eight, or 65.5 percent. The ratio of enslaved children to mothers (children 0 to 10 years old per female slaves ages 15 to 45) increased from the already high figure of 1.125 in 1862 to 1.3 in 1874. In other words, by then there were 30 percent more children than fertile-age enslaved women in Campina. The dependency ratio, which compares enslaved individuals younger than 14 or older than 50 to those who were between 15 and 50 years old, went from 1.212 to 1.521, reflecting a substantial increase in the number of children and elderly slaves.[14]

By 1874, enslaved individuals grew up in the bosom of extended family households. Maria Cassange, an 87-year-old enslaved woman probably born in the kingdom of Kassanje, Angola, lived in Campina with her three grown children; Catharina's daughter Josepha, who was 44 in 1874, married Theodoro (30), who was the brother of Dionizia (13) and Maria Libânia (23); all three were children of Libânia (60). Theodora and Josepha had six children: Joaquina (17), Jeronima (15), Idalino (14), Carolina (9), Gonçalo (8), and Simplício (4). Sixty-year-old Libânia, in other words, lived in the same plantation as her three children and her six grandchildren. Three other family groups lived in Campina: Ingrácia (51), with her three children; Ângela (27), with Emiliana (6) and Alphonse (3); and the siblings Antônio (23), Roberto (20), Francisco (10), and Vicente (9), sons of the deceased slave Margarida. In sum, the demography of the plantation was already well in transition toward that of the surrounding free, rural population, which was characterized by high rates of fertility, mortality, and large female-headed households.[15]

Raiol's purchase of Campina in 1874 did not apparently lead to significant changes in the working conditions of the slave crew. The new master was a lawyer, a respected member of the Liberal Party in the state and the national legislatures, and an intellectual who gained prominence with the publication, among other works, of a famous history of Cabanagem titled *Motins Políticos: História dos Principaes Acontecimentos Políticos da Provincia do Pará desde o Anno de 1821 até 1835*. Raiol's rising star shone even brighter when in 1871 he married Maria Victoria Pereira de Chermont, the granddaughter of Antônio Lacerda de Chermont, viscount of Ararí. Three years later he bought Campina, a property he had visited in 1862 while working as an attorney for Godinho's nephew and only heir, Henrique de Araujo Tavares. After earning a substantial income as a public official and collecting an impressive dowry, Raiol probably acquired Campina as a source of stable income, a symbol of social prestige, and a vacation home—more than as a stepping stone to building a sugar empire.[16] Indeed, during the 1880s he devoted himself to politics and was appointed president of the provinces of Ceará, Alagoas, and even São Paulo; his upward trajectory culminated in 1883, when Emperor Dom Pedro II granted him the title of baron of Guajará in recognition of his intellectual and political service to the Imperial state.

The enslaved, therefore, continued molding the plantation's landscapes according to their own agendas of community and autonomy. William Edwards noted that they "had ways of earning money for themselves, and upon holidays or other times received regular wages for their extra labour." Some served as hunters, guides, and rangers not only for the planter but also for the occasional visitors journeying to the Amazonian forest—just as did other enslaved guides as discussed in chapter 3. The "faithful hunters" who accompanied Edwards enjoyed considerable autonomy to navigate the rivers of the Tauapará region, carried firearms, and even took visitors to the nearby city of Vigia.[17] Against the intended "geography of containment" that the masters of Campina devised, the enslaved "modified" the landscape by carving out "an alternative mapping of plantation space," an "insurgent geography" based on mobility and autonomy.[18]

This process was also visible in the slaves' living quarters. On other sugar mills dating from the seventeenth and eighteenth centuries, there were large, "L-shaped" slave barracks— senzalas—each containing fifteen to twenty cubicles and a veranda. However, L-shaped slave barracks only appear in an oral testimony gathered by writer Sylvia Helena Tocantins in the early 1940s.[19] Neither William Edwards's account from 1849 nor the baron's sale deed from 1874 makes mention of them. Instead, Edwards describes "the houses of the blacks, structures made by plastering mud upon latticed frames of wood, and thatched with palm-leaves." Godinho's 1862 postmortem inventory lists "houses in a bad state," and the 1874 sale deed notes "palm-thatched huts."[20] If a senzala ever existed in Campina, it was later replaced by a series of wattle-and-daub huts hosting slave families, in accord with the process of peasantization taking place during the last decades of slavery.[21]

Before purchasing the mill, Raiol had been an active supporter of gradual abolitionism since at least 1869, and as a lawyer, he had been a *curador de escravos*, or slave attorney.[22] In 1877 he began hiring and bringing migrant agricultural workers to Campina, and in an 1884 newspaper article he asked other notable men of Pará to follow his example and free their slaves.[23] Raiol did just that in 1876 when he freed his slave Josepha, receiving 700 milréis in compensation from the state's Slave Emancipation Fund. Josepha's husband Theodoro and their children— Jerônima (25), Gregório (20), Carolina (19), Gonçalo (18), and Simplício (14)—were freed in 1883 and 1884 for similar sums. The Slave Emancipation Fund prioritized the liberation of families over that of individuals, so the slaves' earlier development of family and kinship ties had the effect of accelerating the emancipation process.[24]

In 1888, the abolition of Brazilian slavery signaled the end of legal bondage in the Americas. There was a big party in the city of Vigia on May 14, 1888, but according to a descendant of the Campina slaves, the baron did not want emancipation to be celebrated in Campina. After days of raucous celebrations in the cities, many slaves left their homes and roamed around for a while, enjoying their newly won freedom of movement. As the initial joy evaporated, however, some found that staying on their former plantations was a safer course of action than venturing out into a world full of uncertainties. At least, in the old plantations they could normally perform their former tasks in exchange for a salary. As their Brazilian counterparts did, a number of Raiol's slaves stayed on the property as wageworkers, eventually reworking the landscape to create full-fledged black peasant communities.[25]

Landscapes of Post-Slavery Freedom, 1888–1940s

Domingos Antônio Raiol was sorely disappointed when Brazilian elites sent Emperor Dom Pedro II into exile and proclaimed the First Republic in 1891. With the emperor's departure, Raiol decided to retire from formal parliamentary politics, embracing his intellectual activities and spending longer periods of time in Campina. Between 1890 and 1894, he wrote the last three volumes of *Motins Políticos*, cementing his reputation as a respected intellectual. In 1900, he was also among the founding members of both the Historical and Geographic Institute of Pará and the Paraense Academy of Letters.[26]

After the baron's death in 1912, Campina continued to function as a sugar mill, but nobody in his family seemed to be very concerned with supervising it during that decade and the next. That was true certainly of the baron's sons. José and Pedro Pereira de Chermont Raiol each received one-half of the property in their father's testament, but they focused on their professional careers in Rio and São Paulo, not on agricultural activities. The ownership of Campina was transferred back and forth in the wills of the baron, José, and the baroness, ending up in 1925 entirely in the hands of Pedro, who sold it to Vigiense rural proprietor Plínio Campos in 1928.[27]

The freedmen of Campina stayed on the property as resident mill employees, agricultural tenants, and even supervisors. Former Cacauense Ana Maria dos Santos remembers that her grandfather Gregório, son of the slaves Josepha and Theodoro, acquired the name of Gregório Moraes in 1888 and became an overseer at the sugar mill. He was not only paid a cash salary but was also given a land plot where he grew crops and lived for decades, along with his sisters Carolina and Jerônima, his wife Maria da Luz, and children. When ownership of Campina switched hands from the Raiols to Campos in 1928, Gregório and his family relocated to Vigia, the city across the Bay of Guajará. Thus, by maintaining his residence under a favorable arrangement, a former slave was able to become a supervisor in the sugar mill.[28]

Several freedmen continued to work at the mill while cultivating a land plot and collecting forest and mangrove products—which most likely in this period was already considered a de facto right for them. Such was the case of Dona Guilhermina da Conceição Goulart and her brother Seu Nunhes, born in 1916 and 1926 respectively, in Mané João—a sítio close to the mill where their parents Irineu and Theodora, the daughter of freedman Antônio, had moved a few years earlier. Dona Guilhermina and Seu Nunhes were wageworkers for Plínio Campos, until he finally dismantled the mill and sold it by pieces in 1938. Seu Nunhes explained to me how they went "to the cane field, cutting cane, shouldering it, washing it, and bringing it to the sugarmill," where it was processed to become sugar, cane brandy, and molasses. The produce was still exported to Vigia and Belém, but unlike under slavery, the workers now received daily pay for their work: "one *tostão*, 200 réis, one *vintém*, a very low salary." He also recalled how the hydraulic system of the water mill was used to propel large barges out of the Campina stream. Visitors "embarked in the boat when they were going to leave. . . . They sat there, the blacks raised the lock gate, the water came out, and they moved away, away. . . . It was a very well-constructed thing." The group that settled in Mané João eventually grew in number, forming the community known currently as Cacau.[29]

Other freedmen occupied a homestead with their families on the outskirts of Campina during those years. Freedman Simplício, in his thirties by 1900, established himself in a community known as Terra Amarela. Freedwoman Carlota went to live in Figênia accompanied by her siblings Andreza and João. Gonçalo, another brother of Gregório, moved to a sítio called São José, and Vicência, Vitorina and Nicolao, Leopoldina's children, went to live in Pedreira.[30] They formed communities of varying sizes, some of which, such as Santo Antônio do Tauapará and Terra Amarela, are still in existence; in the process, they became practically free from any supervision by the legal owners of Campina at some point around 1900.

While the arrangements made with proprietors and managers are difficult to ascertain, it seems that these autonomous black farmers became *roceiros*, cultivators who settled on unused lands peripheral to Campina in the hopes of becoming proprietors by custom. "No, sir, we lived there and worked for ourselves," repeated Seu Alcides, whose grandparents lived in Ovos, when asked whether they had to pay rent for residing there.[31] When new owners purchased Campina, such as Plínio Campos in 1928 or cattle rancher Francisco de Mello in 1938, they probably considered that the marginal lands the farmers occupied, and the length of time these groups had been settled there made it inadvisable either to request tenancy payments or to sue them as illegal occupants.

As during the last decades of slavery, the family was the basic productive and social unit of the community's landscapes. And as under slavery, women represented the stable head of the households, to the point that it was common to find children from different fathers living under the same roof.[32] Family units also structured each village's economy, which rested on three pillars: agriculture, forest collection, and wage labor. Manioc gardens, most often worked by women, formed the basis of subsistence, along with corn and beans, which could also be sold to regatões. The Cacauenses also cultivated yams, rice, okra, pumpkins, and watermelon.[33] Forest items such as fruits, timber, rubber, drugs, vegetable fibers, and nuts, were used either to enrich the villagers' diet or for petty trade. Fishing and hunting in the forest and the mangrove, usually masculine occupations, provided lowland *pacas* (*Cuniculus paca*), agoutis, spider monkeys, turtles, *turú* worms (*Teredo navalis*), river crabs, and fish. Finally, the Tauaparaenses also engaged in wagework—the third pillar of the freedpeople's conomy.[34] In Terra Amarela, for example, Seu Alcides's grandfather worked as a blacksmith for the owners of Campina, and Seu Santana's grandmother worked as a *braçal*, or occasional agricultural worker. For these tasks workers received a daily salary or *diária*, and usually, they did not sign a formal contract, a characteristic arrangement in other regions of Brazil during the post-emancipation years. Ex-slaves probably wished to preserve their autonomy by avoiding long-term contracts, and ex-masters did not want to commit themselves to providing stable salaries.[35]

Community labor complemented that provided by the family units, as the production of manioc flour implied a series of tasks that were best done collectively. At the beginning of the rainy season the household adults cut down and burned the standing vegetation on a two- to three-acre plot of forested land to fertilize the soil (see figure 5). Then came the planting of the manioc stems, usually organized by women with the help of a *puxirum*—a labor drive provided by the community. The harvest came ten to eleven months later, and the same field was then replanted between three and five more times before moving to a new one. The manioc was boiled, grated, and toasted in a manioc oven, a kiln with a large copper or iron plate on top (see figure 6). Living in a community instead of in a single-family *sítio* or farm facilitated carrying out the puxirum and pooling resources to build and maintain the manioc ovens.[36]

In addition to its practical utility the puxirum represented an opportunity to celebrate the black peasants' shared African ancestry. After clearing the fields, the hosts would distribute special drinks and foods like *aluá* and *manicuera* among the participants, a signal that work was over and a party had started. Manicuera is a type of *mingau* or porridge made with rice and manioc flour, and aluá is an alcoholic drink made of pineapple rind and cane liquor that orig-

FIGURES 5 AND 6 Roasting farinha in the 1870s and in 2009. Smith, *Brazil, the Amazons and the Coast*, 383, and author's photograph in Cacau. Courtesy of University of North Carolina State University Libraries.

inated in western Angola.[37] As plates and cups were emptied, some Taua-paraenses made a circle, sat on two or three large drums built from hollow trunks, grabbed a guitar and other instruments, and started singing and clap-ping. Then they took turns to dance in the center, as the rest of them sang. In a syncopated 2/4 beat, the participants responded to the lead singer's calls glossing love, work, nature, and the past of slavery in Tauapará. Named after the drum, the music style these men and women played at the puxirum and other festivities is known as carimbó, the most famous folkloric music genre of Pará.[38]

Carimbó is associated with the fishing regions around the Amazon estu-ary, but there is substantial evidence that its roots run deep among the descendants of the Campina slaves. In 1846, Paraense bishop José Afonso de Morais Torres had already heard how "Godinho's slaves . . . sang to the rhythm of the oars" while transporting him to Campina, and in 1878 and 1879 local newspapers in nearby Vigia had made explicit reference to a music style called carimbó, noting that it was named after the drums used to play it. In 1883, Vigia's Municipal Code of Laws prohibited playing *"corimbó"* music, and on May 13, 1888, the nationwide abolition of slavery was celebrated in Vigia and in Campina with raucous parties where this musical genre figured prominently.[39]

By the 1910s or 1920s, the blacks of the community of Cacau formed what is recognized by most as the first carimbó band in the region: Tauapará Zimba. In 1938, when freedman Gregório Moraes left Cacau for Vigia, his daughter Ma-ria dos Santos Rodrigues became one of the famous *tias do carimbó*, women who organized, promoted, and led the carimbó shows and were recognized as community leaders for doing so. While the word "carimbó" comes from the Tupi word *korimbó* and means hollow trunk (*curi*, stick; *m'bo*, emptied), it is only in the area around Vigia that this type of music is also known as *zimba*, a term of African origin most likely related to samba. More broadly, Brazilian folklorists have established that carimbó has a number of characteristics in common with other styles of Afro-Brazilian music and dance, such as *batuque* and *lundum*. In the city of Vigia there has never been any doubt that Tauapará was one of the birthplaces of carimbó.[40]

The lyrics of carimbó songs provide a unique window into the collective cultural referents of the Tauaparaenses during the first half of the twentieth century. Dona Nadi, from Terra Amarela, recalls how when the inhabitants of Cacau started playing the drums, "A tremendous tremor was felt . . . when they started playing, we said, 'Look, the carimbó is starting in Cacau, let's go there!' and oh God, we stayed there until sunrise."[41] The carimbozeiros frequently portrayed their work on the sea, the *roça* (the farm), or in the forest in songs such as "My Ship is Leaving," "Row, Oarsman, Row," "Mom Treads on the Corn," "I Want to Collect Açaí," or "Seringal [rubber field]." To subvert the hardships of agrarian labor, the black farmers practiced "the deeply pleasurable, healing movements of dance" as they used their bodies to make "the gestures to accompany the [job described in the] music," recalled Dominga Moraes in 1977. Naturally, songs about love and leisure were always present. For example, references to the universal and sexualized *morena*, so omnipresent in Brazilian popular music, were frequent.[42]

The lyrics of the Tauaparaenses' carimbó songs also negate the idea that their social memories of slavery had been forgotten after emancipation. Just as did their maroon descendant counterparts from the Trombetas River, the Tauaparaenses used images and metaphors of nature to frame a series of narratives denouncing the horrors of the slave past. In "The Sugarcane Fields in Flames," for example, the unknown composer explains how "a sugarcane field got aflame" and asks a female partner to "call more people to put it out." But the composer also states, "I am not going there / I'm scared / the fire can catch me." For carimbó participants in Tauapará, the sugarcane field embodied the experiences of past abuse and death suffered under slavery—experiences far worse than fire. It represented, in this specific context, a "mnemonic," an "active agent of identity" with "real material and ideological effects on persons and social relations."[43]

The composition "Negro in the Trunk" described more explicitly the experience of slavery. It tells the story of a "negro" who "came back sweating / from the sugarcane field / the Master called the overseer / and the negro was thumped." In addition to mentioning a *nego* or negro (a name generally applied only to slaves and ex-slaves), a master, and an overseer, the author refers to a specific place of great meaning to the enslaved: the "tronco" where "the Master had him [the slave] tied / with chains around his feet." The song ends by picturing the "white man's wife coming / weeping like a child / don't beat the old black / he is a trustworthy man." The song refers to two generic, faceless enforcers of the slave regime, the slaveowner and the overseer, along with an old black man considered "trustworthy" by the composer. By pitting a slave master against a well-known stereotypical figure, the author was emphasizing the emotional and physical wounds that slavery inflicted on the inhabitants of Tauapará. To anyone living in Vigia, the flogging trunk of Campina is clearly an "enduring memorial" dedicated to the ancestors who suffered the whip; it is a "point of reference" for the shared narrative of slavery as a system of class and racial terror.[44]

From a different perspective, the tronco that appears in the song "Negro in the Trunk" also anchors the changes that took place in the Tauapará landscapes as free black peasants replaced slaves. Until 1888, the captives lived near that heinous display of the master class's coercive power. In later decades, however, the freedpeople gradually dispersed across the periphery of the property to form autonomous farms and communities, and the trunk gradually became a symbol of the black peasants' shared ties to the place. "The reorganization of a landscape" on the island of Tauapará during the early to mid-twentieth century indicated "a realignment of authority," the passage from a "geography of containment" to one characterized by the autonomy and self-determination of a peasant livelihood. While this process took place mainly in the decades after abolition, it was "based on existing communal ties and economic practices forged under slavery," a fact periodically recognized and celebrated by the Tauaparaenses.[45]

Citizens: Political and Juridical Dimensions of a Landscape

Let us now take a look at how the Tauaparaenses tied their experiences in the landscape to their ideas about citizenship. When new landowners such as Campos and Mello arrived in Campina in the early twentieth century, they were initially unaware of the customary arrangements between the black farmers and the former owners. Those newly arrived landowners tried to charge the blacks rent as if they were tenants, and the Tauaparaenses naturally refused to pay, alleging instead that they were "citizens of Tauapará." This expression refers to transformations not only in the landscape, but also in the political arena. In this section, I analyze three relational spheres that shaped the emergence of this discourse in the community of Cacau: the black peasants' relations with the legal world, their ties to political patrons, and their dialogues with other peasants.

The juridical dimension of this narrative started with a document whose exact origin remains unclear. Guilhermina da Conceição Goulart, the granddaughter of the slave Antônio, told me that, before the baron's death in 1912, a man called Raimundo "went there [to Campina] to make a map for the owner. Raimundo then gave it to a man called Plínio Campos."[46] In interviews held in 2009, Seu Zacarias, born in 1943 in another community, and Seu Alcides, a Cacauense born in 1940, mentioned that a posse deed had been granted to the inhabitants of Cacau in 1908. Seu Zacarias, Seu Alcides, and Dona Guilhermina were all referring to the same document: "the map," a posse deed listing all the families of slave descendants who lived on the property of Campina in 1908.[47]

We do not know exactly why and how this document was created. Perhaps the baron agreed to grant the lands of Cacau to his former slaves, although this makes little sense given that he bequeathed Campina to his son. Another possibility is that the baron gave his former slaves this posse deed, which his son Pedro and the baroness simply ignored. Or perhaps the baron gave the slaves the deed of a sítio, and this information was lost in the multiple sales of the estate. Still another hypothesis is that the "map" was in reality a simple census or list of residents, crafted with an unknown purpose. But whatever its origins, this document was not important because of its legal value. Rather, it represented the Cacauenses' first legal victory—however precarious—in their quest to appropriate the lands of Cacau. It carried, in other words, much symbolic weight: it made the Cacauenses appear not only as legitimate heirs of the baron, but also as state-recognized owners of the lands they inhabited.

The Cacauenses' encounters with the landowners of Campina conferred a new layer of legitimacy to their claims. A cattle rancher from northeastern Brazil, Francisco Mello bought Campina in 1938 and, according to Dona Guilhermina, requested that the freedmen's families work for him.[48] In making that request, Mello reportedly used the 1908 map of inhabitants to determine who was living there. Sylvia Helena Tocantins, Mello's daughter, related in her memoirs that her father employed some black peasants who "had lived there during slavery." They included Theodora (Dona Guilhermina's mother), Juvenal (a former slave who became Sylvia Helena's main informant), and many others. Mello "did not want people from outside Tauapará to work the land, but rather the ones with roots," so apparently he valued the slave descendants' attachment to the estate.[49] That he also settled families of refugees from northeastern Brazil fleeing drought in the different sítios of Campina suggests a likely preference for employing families instead of single men.[50] This would be an intelligent strategy in the Amazon, where it was difficult to find and maintain a stable workforce.[51]

Ilson Pereira de Melo (Seu Cebola) is Dona Guilhermina's son, born in 1946. His recollection of Mello's words reflects important aspects of the Cacauenses' interaction with the landowner: "at that time [Mello] said 'ask them [those who had left Campina] to come back, because this is their birth cradle. Nobody has to leave' . . . this belongs to the blacks born and raised here. If the owner wants to sell, then he needs to tell the new owner 'look, you have the map here: here they are. Here they have their families, egg and flour, man and woman, so they should live here.'"[52] That Mello wanted the slave descendants to keep on living in Campina is corroborated by Mello's daughter and by the oldest living Tauaparaenses: Dona Guilhermina and her brother

Seu Nunhes, who was born in 1926 in Cacau. But regardless of what Mello actually said, his appeal is remembered as a claim to unrestricted access to the lands of Campina for the descendants of the slaves who continued to live there, and even for those who had left. Mello must have had his own plans for Campina, but Seu Cebola appropriated his words and transformed them into a legitimation of the Cacauenses' strategies. Ultimately, then, the rancher's words were interpreted as adding clout to the black posseiros' political agenda.

Matters became complicated after Mello's departure in 1943, when cattle rancher Rodolfo Englehard acquired Campina with the goal of expanding its herds. Initially, Englehard tried to impose a new rent of 30 percent of all forest products gathered there: *turú* worms, crabs, fish, game, and wood.[53] In several interviews, Seu Nunhes narrated how four representatives of the black communities went to visit Mr. Englehard's father Alberto, mayor of Belém, to complain about this alteration of the traditional arrangements:

> We went there, to Belém, to Quadrinho Square, where Alberto Englehard [Rodolfo's father] lived . . . "Colonel, I want to speak to the patron" . . . "Where are you from?" "I'm from Vigia, from the fazenda Santo Antônio da Campina." "Oh, ok." . . . "Colonel, I live there. But your administrator prohibited us to do all that we do to produce something. We have lived there for years, we are born and raised there. How come we should pay? Then, what will I produce?" And do you know what he said? "Yes, but what about the *paca* that you brought? What about the crabs, the manioc flour . . . ? You are living there for free . . ." Yes, that could be, we were living there for free, but how could we pay? We grew up there, taking care of the property for free![54]

While this last sentence was pronounced in a tone of indignation, in reality this statement misstates the relation between Englehard and the Cacauenses. They did not have the responsibility of "watching over" the land. Only old Gregório had done so, given his position as overseer. Seu Nunhes's statement presented their process of territorial appropriation as an action of loyalty, not defiance. The goal of the freedmen was to maintain as good relations as possible with Englehard while preserving their unrestricted use of the land and their access to wage labor. Seu Nunhes continued to explain that, in case they were forced to pay,

> "I will leave, but you will pay compensation . . . look, 42 families of the blacks living there . . . now, think about it, how do you want those families

to pay? No sir, you either liberate us from any payment or you will pay us a compensation for the 42 families." And then I showed him the paper I had in my pocket [a copy of the 1908 "map"] . . . And he said, "Ok. You watch over the place of Mané João [Cacau]." And I said, "Colonel, I can watch over it, but only if you put it on paper." And he said, "Ok, go to Soure and talk to Rodolfo, he will resolve the issue."

Alberto Englehard paid for their trip to Soure, on the island of Marajó, where Rodolfo had his cattle ranch. "I told him, 'Sir, I went to Belém and talked to the colonel, he suggested that I came here to talk to you' . . . 'yes, he called me to discuss the issue. Look, I will allow you to work there for free . . . so don't worry . . . you know, your land there, it is a true paradise!' . . . And I said, 'Of course, sir, of course, I am born and raised there, and so are my mother and my father, and all my family, and we've been watching it over for free, sir' . . . And he said, 'Ok then, you take care of the place ['É, toma aí conta de lá.'].'"

But the Cacauenses, knowing that written documents were fundamental in preventing further problems with new landowners, realized that Englehard's verbal statement was not enough. So Seu Nunhes responded, "No, that's not the way, sir. You give me a stamped document, and I'll do it. Because I'm not here alone, I'm here in the name of the forty families who live there." In the end, he explains, Englehard "gave us a letter" to be shown to the administrator, and renounced the plan of charging rent to the blacks of Cacau. By the 1940s, then, the Cacauenses had engaged in a series of encounters with the successive proprietors of Campina during which the owners had recognized the legitimacy of the peasants' claims.

The third relationship shaping the idea of the peasants as "citizens of Tauapará" was between the Cacauenses and other rural populations. All over Brazil, in the late nineteenth and twentieth centuries a plethora of rural smallholders with and without title to their lands defended customary usage as a valid means of acquiring land. On October 20, 1894, for example, the city council of Colares, on the southern side of the island of Tauapará, sent a letter to the governor requesting that a land possession deed (*título de posse*) presented by a group of farmers not be recorded in the state land registry. The council argued that the lands claimed by Firmo de Nazareth e Silva, Francisco Almeida de Nazareth, and others were not *devolutas* or public, but belonged to the municipal patrimony and therefore could not be sold to private parties.[55] The council also stated that its rights not only "outdid" those of the peasants, but actually "pre-existed" them: the council had been created one month before the peasants had registered their posse on December 6, 1854. Moreover, the municipal patrimony derived from a *sesmaria* or colonial land grant issued before 1824, clearly superseding the claims of the peasants.[56]

Yet the council felt the need to present further arguments to buttress their claim. According to the councilmen, Silva, Nazareth, and the other posseiros "pled effective cultivation . . . stable living on these areas, [and] uncontested possession," an explicit reference to the legal requisites to recognize a posse.[57] As we saw in chapter 4, the posse was a form of usufruct rights based, in the case of rural real estate, on uncontested and effective cultivation and residence for ten years. It did not carry as much legal clout as the legal right of ownership, but some courts did recognize it as a proof of residence. This derived from the Portuguese *Ordenações Philipinas* or Philippine Ordinances, as well as from other laws passed in the late colonial period by the Portuguese.[58] In 1850, three decades after Independence, Brazil's Land Law established that public lands could only be acquired by purchase, although in subsequent articles it clarified that the posse continued to exist, extending the mandatory occupation period from five to ten years. While slightly more restrictive, this law still "allowed the posseiros to legalize their posse" by recording it "in the parish land registry" created for this purpose.[59] In 1891, legal ownership over public lands was transferred from the national to the state governments, producing a new wave of laws and regulations that altered the terms of acquiring land by posse, but this legal figure did not disappear. It was not until 1931, with the passage of Decree 19,924, that squatting was no longer a means of acquiring public lands. However, even after that, long-term residence and cultivation continued to be considered a valid source of posse rights in courts of law throughout Brazil.[60]

During the First Republic (1889–1930), Amazonian peasants who lacked an official land deed tried to protect the lands they had been occupying through the right of posse. In 1905, for example, a lawyer hired by Silvestre Sarmiento and his wife to defend their right to a plot of land in Tujuí, near Vigia, argued that "neither do we need, nor does the law oblige us, to show the deed of the posse occupied by our clients, because even if they did not have legal title to these lands, the right of adverse possession [*usucapião*] assists them." This couple was able to prove that they had obtained the land by donation in 1889 by presenting a number of witnesses—the most frequently used valid proof of residence for customarily occupied land. The argument came up time and again in local courts and in letters to police officers.[61] Sarmiento and his wife were only two among the thousands of Brazilians who relied on the recognition of their posses as a mechanism to access landownership during the First Republic.[62]

The idea that traditional occupation eventually generated property rights must have taken hold in the black communities of Tauapará during the nineteenth century. In fact, a group of free peasants fought a legal battle over landownership with Campina owner Agostinho José Lopes Godinho in the 1850s. The convent of Our Lady of Carmo had traditionally rented to local rural workers a plot of agricultural land adjacent to the Campina plantation called Guajará. In 1857, Godinho acquired it and appealed to the courts for the seizure of the manioc harvests planted there. The local judge ruled in Godinho's favor, ordering the confiscation of the crops and the removal of the peasants, but they refused to leave. A second suit was presented, during which Manoel Amâncio Cardoso, one of the posseiros, argued that he did not leave because he "had been living on those lands for many years," adding that the prior of the convent of Our Lady of the Carmo had given him permission to cultivate manioc there. He had no document proving this, but still "he was cultivating and would continue to do so given that the owners really gave him permission to cultivate." Cardoso also declared that he did not know that the lands belonged now to Godinho; instead, he argued, "they belonged to Our Lady of Carmo."[63] For Cardoso and the other defendants, customary usufruct eventually became the right to hold the land.

Apparently, the posseiros never complied with the sentence mandating that they give up those land plots. Twenty years later the baron of Guajará explained how "thanks to the renters [of the Carmo convent] the place [Guajará] is clean and has some fruit trees planted by those who live there."[64] While they had been sued three decades before as illegitimate occupants of Godinho's lands, by the 1890s the baron saw the posseiros as "renters"—as legitimate residents of Guajará, given their long-standing occupation of the area. A tiny group of smallholding farmers had ultimately succeeded in gaining the right to reside and cultivate their land plots. Their settlement was eventually recognized as a village of the Colares municipality in the late 1930s, after its inhabitants defeated a new attempt at eviction. For the Tauaparaenses, then, there were examples of successful battles fought by posseiros against powerful landowners just next door.

The Guajará posseiros' assertiveness, just as the claims of all those who challenged the land grabs of powerful opponents, were all based on the conviction that protracted residence and usufruct of the land constituted an effective path to property ownership. Their chances of achieving the recognition of this right depended on many factors, some of them extralegal and quite difficult to overturn, such as the influence that powerful landowners exerted over local justices.[65] The discursive figures appearing in these conflicts indicate that the struggles of Brazilian posseiros "helped build . . . a culture of resistance that ensured the posse as the core of a right" to own the land where they lived and worked. Landless, illiterate peasants in rural Brazil during the nineteenth century may have not participated formally in the political system, but their defense of customary access to landownership was a tangible embodiment of their struggles to defend their condition of Brazilian citizens. While they were often denied the right to vote, being owners of the lands they cultivated represented "the bases for the disenfranchised and landless to stake a claim to citizenship," thus becoming ultimately an act of a political nature.[66]

Conclusion

From the Cacauenses' standpoint, cattle rancher Rodolfo Englehard could have never evicted them from their community in 1943, at least not without substantial compensation, "because all the local authorities knew that we were citizens of Tauapará." Their ancestors had resided in different spaces in Campina for generations, integrating its agro-ecological world into their community and transforming it into a part of their history through custom. They changed local landscapes accordingly, infusing numerous sites with "real material and ideological effects on persons and social relations," to the point of framing the Cacauenses' perception of who they were and where they came from.[67] Tightly interwoven with bonds of kin, rituals of community labor, agriculture, and folkloric celebrations such as carimbó jam sessions, the natural world internalized by the Cacauenses through their engagement with local landscapes ultimately represented a conduit for a shared slave-descendant identity. Relationships with the environment bound past and present with a thick web of symbols and practices that reaffirmed the Cacauenses' shared ancestry.

But Seu Nunhes's mighty statement also speaks of how interaction with other segments of Amazonian society framed the black peasants' ideas. The fact that some landowners of Campina permitted the Cacauenses to stay on the lands they had inhabited since slavery, for example, was interpreted as an elite sanctioning of their collective project. A mere list of inhabitants in a posse deed likely produced at some point around 1900 also acquired a life its own. This was not an imagined community, but an imagined document sustaining a community thanks to the might of the law. From other posseiros living on the island of Colares, the Cacauenses acquired the legal understanding that protracted and unopposed residence and cultivation of the land were conducive to the legal recognition of the right of posse, considered equivalent to that of property ownership, even though legally it was not. They interpreted this right not only as inviolable but also as a tangible embodiment of their condition of Brazilian citizens. The Tauaparaenses, in sum, merged their communal past as slave descendants with their legal rights as long-term residents of Cacau to see themselves as "citizens of Tauapará."

In the second half of the twentieth century, new landowners arrived in Campina, and their relationship with the Cacauenses took significant and unexpected turns. Negotiations with new landowners like cattle rancher Antônio Avelar in 1964, or with the Caiçara agribusiness company in 1981, were much more difficult due to their adoption of intensive methods of land cultivation and of modern, impersonal forms of land management. Conflict ensued, and more Cacauenses had to leave the community. By the 1990s, 33 families or around 150 individuals still lived in Cacau—and a smaller number in the communities of Ovos and Terra Amarela. Black political organizations had come to Tauapará by then, promoting the opportunity to obtain legal recognition as a "maroon descendant" community and to demarcate their lands collectively, a feat the Cacauenses finally achieved in 2005.[68] But the new processes of identity-building and political mobilization were not built from scratch; rather, they merged with former political discourses and with memories of past political struggles. Even though more than a half-century had passed since he staked the claim, Seu Nunhes's assertion that they were "citizens of Tauapará" would continue to frame the Cacauenses' agenda of defending their rights as Brazilian citizens.

CHAPTER SIX

The People of the Curuá River

*Black Rural Protest and the Vargas Era in
Amazonia, 1921–1945*

In 1921, the village of Pacoval, located not far from Santarém on the northern
shore of the Amazon River, was in turmoil. On March 12, the city council had
asked the state government to stop the privatization of the Praia Grande Brazil nut grove, "the vastest and largest castanhais on that river [the Curuá], where
most Brazil nut collectors reside during the harvest period."[1] In response, in
August the state government sent a special envoy to ascertain if the purchase
and demarcation of Praia Grande were being done by the book and whether
permitting its privatization was a wise policy.

But as soon he arrived in Alenquer, the governmental envoy Palma Muniz
was warned that his inspection "would perhaps not take place because there
was great determination to prevent it, and if he embarked on the Curuá River,
the . . . inspection visit would end at Pacoval," where there lived maroon descendants who collected Brazil nuts in Praia Grande.[2] A local merchant also
warned Muniz against bringing his surveying instruments "because the people
of the Curuá River will revolt upon seeing them." Two days later, as Muniz finally set out to survey the area, a launch with forty people approached the
vessel cheering "for the governor . . . for Palma Muniz, for the blacks of Pacoval, and for the Curuá River in liberty," trying to obtain a hearing with the commission. "They were the so-called BLACKS OF PACOVAL," explained Muniz.[3]
He continued, "After denouncing some Alenquer merchants for not exclusively
using the blacks of Pacoval in the Brazil nut harvest," the maroon descendants
demanded that the Curuá River became "a sort of patrimony of the blacks, so
that only they could collect nuts" there. Having heard their claims and completed his inspection, Muniz recommended in the report he submitted on the
last day of 1921, that the privatization of the Brazil nut groves proceed as sched-
uled. This would avoid "the invasion of adventurers" who "profited from the
work and sweat of the poor . . . without investing any capital in the lands they
have exploited"; that is, unscrupulous merchants who exploited the black peasants and were, according to Muniz, behind the protests of the Pacovalenses.[4]

But far from being the actions of "adventurer" merchants, these protests,
which were described in the introduction, represented a new episode in the

history of Afro-descendant struggles for citizenship in post-emancipation Brazil. This chapter analyzes the three core elements of these protests. First, the Pacovalenses presented themselves as "the people of the Curuá" River and fought to keep it "free," locating the rights of citizenship yet again in the natural landscape. Second, they tried to protect the networks of economic and political patronage that they had built since the time of slavery, which had provided a precarious but real degree of institutional leverage. Finally, in their encounters with public authorities the black peasants also portrayed themselves as "good Brazilians," a nativist claim that mirrored Afro-Brazilian discourses in other states in those years.[5]

The 1920s were a convulsive decade, in which nationwide protests led to the downfall of the First Republic. In a climate of political tension, military unrest, and economic crisis, a traditional politician from Rio Grande do Sul named Getúlio Vargas seized power in 1930. For most of the First Republic the priorities of the state government had been clear: encouraging agriculture and the colonization of Amazonia by promoting large businesses and fixing the workers to the soil—regardless of the long-term social consequences in the region. Would things be different after the 1930 revolution? The promises of the new federally appointed governor Joaquim Cardoso de Magalhães Barata seemed to indicate so. In what follows, therefore, I also discuss the extent to which the revolutionary administration paid heed to the black peasants' demands.

"The Sentiment of Revolt that Dominates Them": Alenquer, 1921

Located in the Lower Amazon region, Alenquer was established in the eighteenth century as a mission. Located just east of Óbidos, Alenquer did not experience the rubber boom of the nineteenth century: its economy was based instead on the export of plantation-grown and wild cacao, Brazil nuts, and cattle products. In 1921, the municipality had between 10,000 and 15,000 inhabitants, but only one-third lived in the town proper. The rest resided in sítios and villages scattered along surrounding rivers and lakes, where they combined subsistence agriculture based on manioc with the commercialization of extractive products. Around 60 percent of Alenquer's population was defined in the 1920 and 1940 census as "pardos," a racial category that included all rural Amazonians of mixed Indian, European, and African ancestry, known traditionally in Amazonia as caboclos.[6] Blacks and whites were each about 20 percent of the population, attesting to the importance of slavery in Alenquer between the late 1700s and mid-1800s.[7]

The revalorization of the price of Brazil nuts in the aftermath of the 1913 rubber crash strongly threatened the traditional autonomy of Afro-descendants in the Curuá. As we saw in chapter 4 they tried to retain access to the groves by finding new fields, pitting different commercial houses against each other, and ultimately, allying with some merchants. Because these measures only had a limited effect, the black peasants also started to protest collectively against what they perceived as acts of illegitimate dispossession. In January 1917, the inhabitants of Igarapé-Assu, a village located between Óbidos and Alenquer, revolted against Italian merchant João Miléo when he tried to prohibit free nut gathering at the local castanhais, which had been considered "so far public." Miléo had advertised his purchase of the nut groves only "through a land request published in the *Diário Oficial do Estado*" so the peasants were unaware that they had been privatized. In the face of the collectors' revolt, Miléo left for Óbidos in early March while threatening to return "with armed people." Ultimately the coordinated actions of the Alenquer and Óbidos mayors averted bloodshed, and the Italian merchant was apparently allowed to continue his commercial activities in Igarapé-Assu. Alenquer's mayor also asked his counterpart that "as a measure of prudence" they should meet to discuss strategies to prevent future revolts, but not in Igarapé-Assu. The protesters' outrage "has not receded yet," he explained, because "this is a people zealous of its rights."[8]

It was in this context that the city council of Alenquer sent a letter to the state government of Pará requesting that the Praia Grande Brazil nut field not be sold to the commercial house Vallinoto & Co, as described in the beginning of this chapter. Despite the fact that the house had applied for its purchase and demarcation following official procedures, the letter stated, "The feelings of that class [the nut collectors] are running high, because they perceive that this harms their interests and is an act of plunder, even more so in these times of economic crisis." If the Praia Grande castanhal did not remain in the public domain, nut workers would unleash "the sentiment of revolt that dominates them."[9]

Given the precedent set by the 1917 Miléo incident, the state government decided to send a special envoy to analyze the conflict on the ground.[10] The representative designated to hear the testimony of the parties in conflict, inspect the fields in question, and evaluate the legal validity of Vallinoto's application was the civil engineer and eminent jurist João de Palma Muniz. He was well qualified for this assignment: a high-profile member of the Partido Republicano Paraense, Muniz was a specialist on land policy, an active historian, and the president of the Instituto Histórico e Geográphico do Pará. By sending such a respected figure, the state administration sought to resolve the Alenquer conflict in an exemplary fashion, establishing a blueprint for similar disputes over extractive products in the future.[11]

As soon as he set foot on Alenquer, Muniz noticed the agitation and unrest plaguing the city. Just after he arrived on August 22, some locals warned Muniz that he should not carry "any [surveying] instrument at all" during his visit to the nut groves, because "the people of the Curuá River will revolt upon seeing" them, and his inspection "would end at [the black community of] Pacoval." Merchant Anselmo Ferreira also told Muniz that the inspection carried "a risk of life." The night before the expedition left Alenquer, an unidentified individual yelled from a bridge, "There will be no survey!" Two days later, the episode of the launch with forty people approaching Palma Muniz's vessel and demanding a hearing with the commission took place. Aware of the authority the state government had bestowed on him, however, Muniz replied that he would visit Pacoval only after he had gathered information about the castanhais, and he proceeded upriver.[12]

On his way back to Alenquer some days later, Muniz did stop to hear the "blacks of Pacoval." The village was founded in the nineteenth century as a maroon community, just like the ones from the Trombetas River area described in earlier chapters. In addition to using the waterfalls as a defensive buffer and the forest as a pantry, the Pacovalenses wove thick networks of relations with merchants, judges, tavern keepers, farmers, and even those still enslaved; these contacts provided valuable intelligence on the next moves of the authorities.[13] Pacoval had resisted, for example, the forces sent against it as part of a statewide anti-maroon offensive in 1853 and 1854.[14] In 1876, however, 135 of the Curuá mocambeiros turned themselves in to local authorities, in an astonishing reversal of the tradition of resistance against slavery. Apparently, disease had struck their population, and the maroons felt compelled to reach some type of agreement with Luiz de Oliveira Martins, a local Brazil nut merchant. Some mocambeiros eventually returned to Pacoval as free men after being shipped to Belém, but others were reenslaved there.[15] In 1901, the community was elevated to the condition of *povoação* or village, which amounted to recognizing its inhabitants as regular Brazilian citizens. By the time Palma Muniz visited Pacoval in 1921 he described an austere hamlet hosting "a single street, very narrow and scruffy" and "thatched-roof huts, some with daub on latticed frames of wood, less than 35 in number."[16]

The mocambeiros demanded that the Curuá River be declared "free" and its castanhais a public good. But their idea of "freedom" was more than just an abstract invocation to keep the nut groves unregulated. Rather, it was firmly anchored in the pivotal role that the nuts had played in the creation of the Curuá River area as a space of freedom and autonomy since the mid-nineteenth century.[17] The mocambeiros had started collecting forest goods during the 1870s, and perhaps earlier. In 1881, Luiz de Oliveira Martins, the Alenquer merchant who had led the mocambeiros five years earlier to surrender to the local police, listed no fewer than nine maroons as debtors of his commercial house, which exported mainly Brazil nuts—and he was not alone in illegally trading with them.[18] These illicit exchanges had been pivotal to the survival of the maroons, providing them with added protein in their diet, cash in their pockets, and items as necessary as salt, sugar, coffee, cloth, soap, tools, manufactured goods, and weapons.[19]

In the decades after the abolition of slavery, the free extraction of nuts from Praia Grande and other castanhais had underpinned the mocambeiros' independence from landed elites—and their relations with commercial ones. By the early 1900s, some commercial companies were already present in Pacoval, but the degree of coercion they could exert remained limited. Commercial houses operating in the nut groves near Pacoval in the 1920s included Santos Amaral & Co., A. Vallinoto & Co., Levy Dahan & Co., José da Costa Homem and Fernandes Nunes & Co, and J. Nunes & Co.[20] Relationships with these commercial houses may have included mild forms of debt bondage, but that did not stifle competition between different commercial agents, allowing the maroon descendants to pick the best outlet for their nuts. Instead of rejecting market exchanges, in sum, the mocambeiros sought to embrace them in a piecemeal and controlled fashion. Their claim that the Brazil nut groves should remain "free," then, was an attempt to protect their traditional commercial networks, which had been characterized by a high degree of autonomy.

Ever since the 1800s, such networks had permitted the black peasants to establish bonds of *compadrio* (godparenthood) with local merchants that were not only instrumental to their material welfare but also provided a certain degree of institutional support.[21] Luiz de Oliveira Martins is a case in point. A slave-owning cacao planter and livestock rancher, in addition to being a Brazil nut exporter, Martins was the first explorer of the Curuá River, according to Palma Muniz, as well as one of the first active merchants in the area.[22] He played a role in convincing 135 Pacovalenses to turn themselves in to authorities in 1876 and "financed" a number of them, which made the Martins family name appear frequently in Pacoval during the following years—a sign of deference toward a patron.[23] Alenquer's mayor Fulgêncio Simões, who had previously been president of the states of São Paulo and Goiás and had held numerous state offices in Pará, seemed to play a similar role.[24] He claimed to be behind the 1901 elevation of Pacoval to the condition of *povoação* or village, which meant that police and municipal representatives from Alenquer were stationed there; in 1921, he proposed in the Senate "to grant them [the Pacovalenses] the property of the lands they inhabit, thus guaranteeing their future."[25] By establishing bonds of deference with selected individuals from the Alenquer elite, then, the maroon descendants sought to influence local institutions, a useful resource when it came to keeping the castanhais "free."

Thus, in contrast to the experience in the Trombetas River area, where patronage ties harmed the mocambeiros' agenda, as we saw in chapter 4, in the Curuá such bonds seem to have empowered them vis-à-vis state representatives. In 1901, French explorer Otille Coudreau made contact with Alexandre, a Pacovalense who boldly claimed, "I am the governor of Pacoval. I am the government's representative"—a surprising attitude given the maroons' traditional mistrust of the government even after the end of slavery.[26] The recognition of the village as a povoação that same year, brokered by Fulgêncio Simões, was probably interpreted by Alexandre and others as the confirmation of their right to the lands of the Curuá—and as a legitimation of their agenda for the management of the Brazil nut groves. In a bold and assertive tone, Alexandre explained to Otille Coudreau how "we do not accept that anybody comes here to make the law," because "the land is ours[,] and we are free to do all that we want."[27]

The Pacovalenses's demands made on Muniz in 1921 also included that the Curuá River became "a sort of patrimony of the *blacks*, so that only they could collect nuts" there.[28] Condemning the "merchants from Alenquer" who had brought in their own nut extractors from outside the river area, the Pacovalenses were thus demanding that "exclusively the blacks of Pacoval" be employed "in the Brazil nut harvest," preserving the castanhais as a source of income for the locals. A large number of Italian and Portuguese merchants began moving to the Lower Amazon in the early twentieth century. Praia Grande property applicant Antônio Vallinoto, for example, came to Alenquer from San Constantino de Rivello (central Italy) around 1900, acquired a commercial firm in 1910, and soon embarked on expanding its landholdings.[29] By 1918, Vicente Sarubbi, João Ferrari, Carlos Alaggio, Bras Calderaro, Vicente Felizzola, Antônio Calderaro, Honorato Calderaro, and many other Italian merchants had established commercial firms in Óbidos.[30] It was actually difficult to find an Italian family in Óbidos and Alenquer who did not have a member engaged in extracting products such as rubber, nuts, or various vegetal oils.

Thus, black peasants and traditional, Brazilian-born merchants could employ a shared nationalist language to oppose the expansion of the newcomers' businesses. On December 28, 1925, for example, "forty-odd nut workers" from the Trombetas River, nut merchant José Gabriel Guerreiro Junior, surveyor Agrimino Valmont, and local police agent Manoel Roberto de Azevedo Vasconcellos all met at the house of land surveyor José Henrique Diniz. The protest was almost identical to that of the 1921 Alenquer episode, but expressed an anti-foreign stance even more clearly. The ex-maroons emphasized that they had "resided there [in the Trombetas River] for many years" and were "good Brazilians" seeking protection from the abuses of a ruthless Portuguese merchant.[31] He had displayed "a relentless ambition, is the lord of most of the Trombetas River, [and] exploits and mercilessly persecutes those who are poor and have no protection whatsoever." If the state government authorized the land sale he applied for, they argued, "they will be evicted and they will have no other place to work, losing their [land] improvements, and the despair of struggling for life will throw them to an act of sudden violence, which they, as good Brazilians, want to avoid at all costs."[32]

Anti-Italian aggressions also took place in Juruty in 1923.[33] In November 1925, the Italian consul in Pernambuco, Antonio Luzardi, contacted the governor to complain about the threats directed against Italian citizens in Óbidos and Oriximiná. The immigrants explained that "anonymous leaflets" were circulating in Oriximiná "encouraging the people to revolt against us." A Brazilian worker named Raymundo Marques told the authorities how he was contacted by "five unknown men, black in color, armed with rifles," who asked him "if he knew Braz Miléo and Braz Calderaro, both Italians." They offered him money "to keep it secret and show them where the Italians' shops were . . . the attack on those shops would take place late at night to make those people, who should leave Brazil, definitively disappear."[34] A few weeks later, in mid-December, Oriximiná police officer Antonio Machado Imberiba responded that he had taken "the firmest measures to suppress any disturbance of public order" by reinforcing the police presence and investigating the origins of the leaflets.[35] Ultimately, no large-scale xenophobic attacks took place in Oriximiná at this time.

In sum, the Pacovalenses' defense of the "free" castanhais in 1921 was an economic vindication—and much more than that. Unfettered access to the fruits of the Bertholletia tree guaranteed the viability of Pacoval as an autonomous community, kept alive a symbol of freedom enmeshed in the landscape, and permitted the reproduction of political and institutional alliances vital to its survival amid growing commercial pressure. Privatizing the

Brazil nut would force the Pacovalenses not only to relinquish the land-scapes of autonomy they had erected for at least a half-century, but also to sever their symbolic, affective, and even political ties with the Curuá's land-scapes of freedom.

Avoiding an Invasion of Adventurers: Pará's Republican Government's Response

The Palma Muniz commission report was delivered to the state government in February 1922, and its final recommendation was to go ahead with the privatization of the Curuá lands. Auctioning the nut groves, the report argued, would concentrate land tenure in the hands of wealthy landowners. The option of surveying, dividing, and selling smaller lots was not viable either, because doing so would incur expenses too high for the state government to absorb. A last alternative was to keep the groves public, as it had been government policy before 1910, but according to Muniz this possibility would cause grave problems. "Under the false premise of *protecting the poor*," he explained, this option only "guarantee[d] the invasion of adventurers . . . who have profited from the work and sweat of the poor . . . without investing any capital in the lands" containing castanheiras.[36] Given that no documentary proof had been presented showing that the requested castanhais were on privately owned lands, the privatization process could, and would, continue.

Traditionally, Brazilian elites always saw agriculture on privately owned lands as a preferable option than forest extractivism. From the time Portugal first settled Brazil, the extraction of forest products in Amazonia was seen as a source of economic instability, geopolitical insecurity, and uncivilized habits.[37] True, at the height of the rubber boom, when huge revenues were flowing to the state government, rubber was seen as the hen that laid the golden eggs, and critiques of the extractive economy receded for a while. However, in 1913, the price of Amazonian rubber collapsed, causing a tremendous shock that devastated the state's economy. As rubber exports plummeted, criticisms of extractive activities regained strength: instead of roaming the forests in search of easy money, it was argued, Paraense peasants should grow crops on the land, which would eventually contribute to their stability and prosperity. Failure to promote the stable cultivation of rubber had thrown Amazonia into misery. This was the lesson to be learned, and land policy would be designed accordingly.[38]

Republican politician Lauro Sodré epitomized this position. A former state governor in the 1890s, he ran unsuccessfully for the national presidency in 1898 and then returned to Pará, where he regained the provincial presidential office between 1917 and 1921. In 1919, he celebrated the recovery of agricultural production and the end of rubber's "natural monopoly" on the state's exports. According to Sodré, that monopoly had caused a "failure of our agriculture," depriving it of "its indispensable instruments to live"—that is, laboring hands— and had forced Pará to "import all the necessary to feed the rich and the poor" as well.[39] Other Republican policy makers held similar ideas. In 1925, for example, state president Dionysio Bentes harshly criticized the collection of forest goods: "one of the causes of our belated economic development is the people's interests . . . they do not take root [in the land] through the trees they plant, but instead through their desire for easy money." In so doing they became "nomadic masses . . . living in improvised huts or without shelter, inadequately fed, malnourished, in unsanitary conditions, and almost always stricken by serious diseases that decimate them."[40]

Henrique Santa Rosa, organizer and director of the State Land Bureau during most of the First Republic, was like Sodré a founder of Pará's Republican Club.[41] He shared Lauro's concern with promoting the stable, orderly occupation of the land, and stressed the idea that unplanned settlement—the spontaneous colonization of poor peasants in unoccupied lands—destroyed natural resources. This was clear in the case of lumber but also applied to rubber, because its trunk had to be cut in order to drain the latex, and to the Brazil nut tree, which eventually ceased producing nuts if surrounding vegetation was cleared.[42] "The exploitation of products through free invasion [of the forest] is not only negligent but also destroys invaded areas and occupies them with no permanent settlement," Santa Rosa argued in 1925, echoing the sentiments of Palma Muniz. "The invaders and destroyers of the forest are those who do not want to subject themselves to the rule of the law. They seek the greatest gain with the least investment, paying the local workers miserably."[43]

But while "disinterest" and "nomadic habits" characterized extractivism, "the same is not observed in lands acquired through a property deed, which makes the proprietor feel attached to the lands from which he benefits." For Santa Rosa, the holder of a definitive title considered land "an ingredient of his prosperity, improving it, making it produce, defending its limits, and enlarging it." The argument that granting land deeds to agricultural developers strengthened the economy per se reflected the influence of positivism on first-generation republican politicians in Brazil.[44] Palma Muniz, who shared Sodré and Santa Rosa's political background, also expressed in the 1921 Alenquer report his belief that the "institution of property in the castanhais" would allow the state government to achieve the "progress and growth of its population" and to "defend its riches and its economic sources."[45]

The attitudes of Muniz, Sodré, and Santa Rosa toward social problems also embodied the principles of progress and order, a positivist motto that infused Brazilian republicanism—and that decorated the Brazilian national flag.[46] *Progress*, in this case, meant regularizing land tenure to encourage private initiative, no more and no less: "Private initiative contributes greatly to put our lands to work, investigating their riches and complementing governmental action, which, in turn, will always try to resolve the problems of transportation infrastructure." Republican progress also meant supporting economic entrepreneurship by developing "railway lines, telegraph service (and later telephone service), schools, roads, paved streets, electric power, public light, clean water," warships, and other technologies of the time—rather than more inclusive state institutions or social policies targeted at popular groups.[47]

Order, on the other hand, meant preserving traditional hierarchies. Being "essentially conservative," Brazilian republicanism "took on the character of embracing inequality and of sanctioning the law of the strongest," thus marginalizing popular groups from the decision-making process. Thus, for Muniz, Santa Rosa, and Sodré, the fact that Brazil nut lands were public patrimony in the past was the cause of the reckless exploitation of natural resources, the nomadic habits of the Amazonian population, the poverty of the region, and the lack of revenue for the state's coffers. They wanted to see large businesses practicing a stable agriculture that fed the export trade and encouraged permanent settlement. The black farmers' request to maintain unmediated access to the castanhais was considered wasteful and even harmful; they saw their demands as a problem to be solved either through charity, by reserving small lots of land for common use, or through police intervention—but nothing else. In the words of Washington Luís, the last republican president of Brazil between 1926 and 1930, "the social question is a case for the police."[48]

These principles translated into a series of land policies that, far from resolving agrarian conflicts between the haves and have-nots, exacerbated the inequality in access to land tenure between regional elites and smallholding or landless peasants. First came the leasing of Brazil nut fields to specific individuals in exchange for paying annual rent to the state government. This practice was not as prevalent in the Lower Amazon as it was in the Tocantins River area. It was enacted into law in 1920, expanding greatly during the rest of the decade and becoming ultimately one of the most visible forms of illicit enrichment, political clientelism, and elite abuse during the decade. Nepotism with nut grove leases fueled the ascent of authentic regional oligarchs, such as

Deodoro de Mendonça in the Tocantins, José Júlio de Andrade in the Jari River, and others.[49]

Second, to "propel the economic development" of the state through the "economic exploitation" of its public lands, state governments in this decade granted enormous land concessions to individual entrepreneurs, especially during Dionysio Bentes's administration from 1925 to 1929.[50] The concessionaires were given the right to explore the area, settle colonists, and start mining, agricultural, and ranching activities. In July 1925, Governor Bentes granted the sympathetic newspaper editor José Miguel Pernambuco Filho a concession of 17,424 hectares in Juruti, and gave other individuals concessions of 19,000, 25,000, and 100,000 ha in the municipalities of Montenegro and São Domingos de Boa Vista. By the end of his administration in 1929, the governor had granted concessions in similar conditions for more than nine million hectares, or 21.25 million acres. Many were given to foreign mining and agribusiness companies, such as the million hectares granted in 1929 to Japanese entrepreneur Hachiro Fukuhara in Monte Alegre, or the famous concession to Henry Ford's company in 1926 with the intent of erecting Fordlândia, a "slice of twentieth century civilization" in Amazonia.[51]

In sum, during the 1920s state elites perceived popular demands as a nuisance at best—and as a threat at worst. The state government embarked on the republican project of stimulating agriculture through the promotion of orderly settlement and the protection of the interests of propertied elites. As part of this agenda, the pernicious "nomadic masses" were prevented from enjoying any autonomous access to the riches of the forest. Perfectly aligned with these goals, the 1921 Muniz commission recommendation in Alenquer became a cornerstone of Paraense public policy on natural resources management, because it supported the sale of Brazil nut groves previously considered public to commercial houses. The privatization of Brazil nut groves proceeded apace, causing violent incidents of popular unrest in Baixo Amazonas in 1921, 1923, 1925, 1926, 1927, and 1929. The social and political model of the republic was under severe strain.

A Revolution Was Needed in This Country: The Vargas Era in Amazonia

In 1930, a coalition of traditional and emerging political forces took power in the federal capital of Rio de Janeiro. Led by Getúlio Vargas, a man of the oligarchy from southern Brazil, the new government comprised members of the republican political elite from states marginalized by the rule of São Paulo and

Minas Gerais, young nationalist army officers, and an urban middle class of liberal professionals and white-collar workers. Some sectors of organized labor also supported the Liberal Alliance, as this heterogeneous coalition was called. More than having a clear-cut agenda, the Alliance had a reformist program focused on political regeneration through a broadening of the suffrage, the eradication of corruption and clientelism, the formation of a stronger national government, and a new sensitivity toward the harms of the draconian inequality that characterized Brazil.[52]

Seeking to increase the presence of the federal government in state politics and to curb the power of traditional state oligarchies, Vargas appointed a series of "state interventors" to replace state governors.[53] In Pará, the chosen one was Joaquim Cardoso de Magalhães Barata, a military officer born in Belém and raised in Monte Alegre (Lower Amazon), who had participated in the unsuccessful 1924 coup against the republican government in Manaus.[54] Soon after assuming power on November 12, 1930, Barata implemented a program of reforms based on the agenda of the Aliança Liberal. He strengthened the control of the state government over municipal councils and implemented a series of social and economic reforms designed to ameliorate the condition of the poor. He enacted labor legislation such as workplace compensation and salary increases for municipal workers, imposed rent controls on urban areas, granted access to legal representation for poor people, and revised requirements for a number of public jobs to ensure that Brazilian citizens instead of foreigners held them.[55]

"Our caboclos," argued a journalist sympathetic to Barata, also "need protection." While the rural dwellers of Amazonia "are Brazilians and have the same blood" as their urban counterparts, the rural poor lived "like slaves."[56] The interventor responded that he had not forgotten his "unfortunate patricians inhabiting the vast interior of our country." He regularly took extensive trips throughout the interior of Pará, becoming the first governor to do so in decades. On March 12, 1931, Barata passed a key decree seeking "to protect the proletarian classes, long oppressed by the men of the government and by laws that, instead of favoring them in their work, suffocate them with high taxes on the acquisition of small plots of land, while granting for free thousands and millions of hectares of public land to businessmen favored by political deals." The decree exempted smallholders from paying taxes and fees when applying for official land deeds, "because they have no resources" to afford them. It legalized the grant of small plots of 25 hectares to poor families for free, and established a new mechanism to create agricultural colonies with free lots for unemployed workers, fishermen, retired military men, and officers of the disbanded state police. "A revolution was needed" in the countryside, the interventor observed shortly afterward. Two weeks after approving the decree, the government started granting the free lots.[57]

In a public display of nationalism, the new administration also claimed that it "did not forget the humble, the poor, the unhappy" who had been marginalized to favor foreigners. Thus, Barata attended to a number of letters sent to him by Brazilian workers of public agencies requesting that foreign individuals be fired from their jobs and replaced by Brazilian citizens. "The public powers cannot and should not include among its servants . . . foreigners who should naturalize as Brazilian citizens," explained the interventor in response to a petition to fire Italian citizen Angelo Panzi. Barata often forwarded petitions like that of José Leite, who requested to be named assistant gardener at Republic Square, thereby replacing "a foreigner," to the local mayor or chief of police. On November 24, 1931, the interventor also increased by 50 percent the taxes for all private companies whose workforce was currently not at least made up of two-thirds of Brazilian workers.[58]

Barata also exhibited a markedly personalist and paternalistic style, often micromanaging the peasants' requests directly and without intermediaries. He instituted three public audiences at Belém's presidential palace every week and, as mentioned, responded personally to the large numbers of letters sent by humble citizens from all over the state, often forwarding them to the proper administrative agency. Using remarkably paternalistic rhetoric, he embraced the "poor people" of the interior and overtly admitted that he was "in favor of dictatorship" because "the people has no civic preparation yet." By treating the situation of popular groups "with affection," Barata hoped to resolve social problems while avoiding the autonomous action of the working classes. All in all, he took a classic populist approach to politics, which was also employed by figures like Getúlio Vargas or Juan Domingo Perón.[59]

Barata also revoked the large concessions granted by Governor Bentes to cattle ranchers and agribusiness companies because they "did not advance the interests of the collectivity and seriously harmed the public good." The rationale for executing the legal annulment of the concessions was that they had either "expired" or the contractors "had not fulfilled the[ir] legal obligations," as stated in the decrees formalizing the cancellation.[60] Between January and April 1931, Barata revoked concessions totaling about nine million hectares (twenty-two million acres) granted to proprietors of different nationalities in areas like the Gurupy River and at the border with Maranhão, Amapá, and French Guiana.[61] "The Land Bureau will evaluate the damage done by the contractor to the forests . . . destroyed without governmental permission" threat-

FIGURE 7 Magalhães Barata visiting Henry Ford's Fordlândia in March, 1931. Barata is in the front row, fourth from the right, in uniform. "Impressões do sr. interventor federal sobre as grandes realizações Ford na Tapajonia," *Folha do Norte*, March 25, 1931. Courtesy of Biblioteca Pública Arthur Vianna, Fundação Cultural do Pará.

ened one of the decrees. On March 13, the Land Bureau's chief engineer Jorge Hurley was appointed to form a commission "verifying not only concessions that are either expired or should be cancelled, but land demarcations granted" before the Bentes administration as well.[62] Henry Ford's tropical paradise of Fordlândia was an exception. A visit to the company's facilities in March 1931 (see figure 7), convinced Barata that the company town was "a center irradiating labor, occupying thousands of Brazilian hands" and contributing "to the interests of Pará and of the Brazilian nation."[63]

Most importantly, the interventor also undertook a revision of all the laws regulating the extraction of Brazil nuts and other forest items—the activities that had led the black peasants to revolt during the former decade. Catering to the requests of some poor extractors, Barata did not hesitate to move against the excesses committed by "patrons who exploited the extractors"—"feudal lords" such as Almeirim's José Júlio de Andrade, Marabá's Deodoro de Mendonça, or Souzel's Porphirio de Miranda. In addition, he established that, to gain more power over the system of concessions of Brazil nut groves created in 1920, the state government would rent the lands with extractive resources only on a year-to-year basis. Finally, the administration created the Bureau of Mines and Castanhais and appointed a commission to revise the previous privatizations of Brazil nut lands, especially the abusive "perpetual leases" created in 1925.[64]

This commission seemed surprisingly vigilant in overseeing, stopping, and even reversing the land acquisitions of some prominent Lower Amazon nut merchants. In June 1931, it halted the purchase of Oriximiná's Santa Rosa property, requested by José Antonio Picanço Diniz, though the sale was eventually authorized because the application process was found to be in order. Others did not have the same luck. Raimundo da Costa Lima's application for a property on the Craval River was annulled by the governor in 1933, as was Costa Lima's attempted purchase of the Tres Barracas property in May of the same year.[65] The complaint presented by Clemente da Silva and other extractors against Italian merchant João Miléo in December 1931 led to an investigation conducted by the State Land Bureau of the accuracy of his land purchases. During a trip to Óbidos and Alenquer that month, Barata heard the stories of the nut extractors in person and then directed the newly created public service of pro bono legal counselors to settle a number of conflicts between extractors, merchants, and landowners. The intervenor even had Italian merchant Braz Miléo arrested for his "disrespectful attitude and the threats he makes to the government by invoking the name of Mussolini."[66] For the first time, the big Brazil nut merchants of the Lower Amazon were encountering a state government that actually supervised land transactions to enforce the law and avoid abuses.

The most telling example of the shift in the governmental position regarding agrarian conflicts over Brazil nuts is the so-called Paiol case. In 1927, the public was shocked when, in the course of a protest against the privatization of some castanhais in Óbidos, store manager Antonio Barroso Pereira and his clerk João Leite da Cunha were killed.[67] The police eventually captured most of the rebel extractors, except for their leader Santa Rosa, who escaped to the forest. They were tried for murder and acquitted by a local jury in late 1929, on the grounds that it could not be established which of the fourteen defendants had shot Leite. The public attorney appealed, and by late 1931, when Barata visited the Óbidos jail, the defendants were still behind bars awaiting a retrial.[68]

Barata's assessment of the case could not be more different from those of state administrators in the former decade. While in 1927 the Paiol rebels were called "bandits" and "beasts" in the press, he proclaimed that the residents of Paiol "were evicted by bullets from the lands from which they and their families drew their subsistence, and saw their houses burnt down by those who in those days could do anything [they wanted]." Catering to their demands, Barata bombastically promised reporters that there would be a prompt review of their case. In the three years left of his administration, however, such a review did not happen.[69]

Obviously, it would not be easy to reverse the power of local oligarchies, which had dominated the Brazil nut trade for almost two decades. It would require a sustained effort with solid legal and political foundations, coupled with a viable alternative to the local oligarchies' power. Unfortunately, Barata had neither. In his four years in power he had alienated a number of powerful men, such as Porphirio de Miranda or José Júlio de Andrade, and was unable to erect durable networks of clientelism with local politicians to displace the most important merchant families.

Take the Guerreiros of Óbidos, for example. While after 1930 they sought to appear as supporters of Barata, in the past they had been bulwarks of the Republican political order through their ties with Lauro Sodré's political machine. Initially, they had supported Barata's reform measures at the local level and obtained important local and state offices in exchange. However, when the old elites overthrew Major Barata, they quickly switched sides and ran to support the new administration. The Guerreiros exemplified the fragility of Barata's political networks: they comprised a number of clans who, just as easy as they got in, could get out.[70]

Eventually, the precariousness of Barata's political machine and his lack of experience navigating the turbulent waters of state politics brought his first administration to an end. Even though his Liberal Party won the 1934 elections to the Constitutional Assembly, when the state assembly met in April 1935 to write a new state constitution and choose a new governor, seven elected representatives of Barata's own party declined to vote for him. In response, he ordered the local militia to occupy the legislature, allowing only loyal representatives to enter the assembly. When the seven dissident legislators tried to enter the building accompanied by a judge and a military escort, a skirmish erupted and shots were fired. Two legislators were wounded and two civilians killed. Shortly afterward, Getúlio Vargas removed Barata from office, and the legislature elected as governor José Carneiro da Gama Malcher from the Frente Unida Paraense, which represented the traditional, anti-Vargas oligarchical forces.[71]

Once Governor Malcher took office in mid-1935, he reversed most of Barata's policies. While Barata had designated some large castanhais for public use, Malcher overturned that action, liberating them for private purchase. Powerful landowner Deodoro de Mendonça, among other republican oligarchs, returned to the forefront of state politics after being ostracized earlier.[72] Most of the land sales under review by the Brazil nuts land sale commission were now granted, including those requested by Costa Lima and other prominent nut merchants from the Lower Amazon. Costa Lima's application to purchase the Tres Barracas property was halted in December 1933, but was granted by the new administration in November 1935, with Barata no longer in power.

Merchant Fernandes Nunhes's request to purchase the "Central" property was similarly stopped in 1932, but was eventually granted in 1938 by the Land Bureau under the Malcher administration.[73] The nut merchants were thus able to turn back the clock to the pre-Barata era.

Conclusion: Black Demands for Citizenship and the Revolution That Was Not

The Pacovalenses' demonstration during Palma Muniz's visit in 1921 and the protests of other black peasants throughout Pará during the rest of the decade were pivotal events in the history of Afro-Amazonians, because they signaled their entrance into the realm of modern mass politics. New as these actions were, however, they were based on a political agenda firmly rooted in the landscapes the peasants had commanded ever since the era of slavery: their idea of freedom for the Curuá River was anchored in the selective usage of commercial networks that the ex-maroons had practiced for decades. Autonomy from landed elites and the pursuit of wage labor had been core to their reproduction as free peasants for generations, and the capacity to freely dispose of Brazil nuts was the only way to maintain that freedom. The metaphor of the "free" river evoked the black peasants' agro-ecological traditions, of which Brazil nuts were the clearest embodiment. The rights to maintain customary access to natural resources and to freely dispose of their labor as they found best lay at the core of Afro-Brazilian notions of citizenship and were expressed using tropes from the natural world.

The Pacovalenses erected commercial and fictive kinship bonds with notable patrons like Fulgêncio Simões or Luis de Oliveira Martins, legitimizing their customary relation to the land and substantiating the idea that it belonged to them. "The land is ours and we are free to do all that we want, we do not accept that anybody comes here to make the law," a maroon descendant of Pacoval argued in 1901.[74] A long history of complex interactions with local elites had cemented a sense of property and the awareness of a specific historical trajectory, distinct from that of other peasant groups. For decades the Afro-descendants of the Curuá River were apparently successful in establishing bonds of collaboration and mutual dependency with local elites, feeling empowered and attaining a real, if precarious, degree of influence over local institutions in exchange for them.

The black peasants also felt threatened by the rise of new ethnic groups that used harsh strategies of labor coercion, which severed traditional commercial arrangements. "My fellow citizens are gradually being evicted from their homes by a syndicate of foreigners, who, protected by their countrymen . . . are acquiring the castanhais in this municipality, evicting the nationals or subjecting them to a state of semi-slavery," Alenquer landowner Luiz Marques Baptista explained in 1935. The mocambeiros responded that, as "good Brazilians," they did not wish to use violence, but if the land grabs of Italian and Portuguese immigrants continued, they would "make those people, who should leave Brazil, definitively disappear."[75]

During the First Republic popular groups all over Brazil generally saw their economic and conditions deteriorate, and the mocambeiros of the Lower Amazon were no exception. Their agency was systematically denied; it was alleged that "behind our unhappy and ignorant fellow citizens, there continues to be the sordid ambition of foreign merchants, who . . . apply the tactics practiced here by the famous Francisco dos Santos Amaral, encouraging them to revolt."[76] The 1921 case established the policies the state government would follow throughout the decade whenever conflicts on Brazil nut lands erupted, leading the maroon descendants to lose their most effective source of access to cash.[77]

Yet their nativist appeals to the authorities and to local elites, and their displays of strength and numbers during the visit of the Palma Muniz commission in 1921, started a new pattern of political action. Patronage surely persisted for a time, but these displays of strength in order to sway institutional action in their favor were qualitatively different from the old deals with local patrons from the 1800s. Amazonian maroon descendants entered the realm of contemporary mass politics during the 1920s—not in 1988, as many researchers specializing in the study of contemporary quilombos argue.

Magalhães Barata was appointed by Getulio Vargas in 1930 to carry out the program of the Liberal Alliance in Pará. With his nationalist stance and his calls to protect poor peasants, he seemed to finally cater to the black peasants' interests. However, he was in power for too short a time to carry out his project of democratizing access to land, and ultimately his policies did not have sufficient political support to outlive his ouster from the presidency. It is likely that, even if he had spent more time in office, the entrenched interests that he confronted and his limited skills as a politician would have hindered his achieving the ambitious goal of facilitating access to land for the have-nots. His executive and personalist style undermined his policies, something that became clear once he left government. A single man could not reverse decades of ill-designed governmental policies—not even from the governor's seat.

Years later, in 1943, Barata was reelected to the state presidency and returned to power in Belém. By then, however, Brazil nuts were no longer the important source of state revenue that they had been in the post-rubber years. The volume of exports decreased tremendously during World War II, and when they grew again after the war, new areas like the Tocantins River, and Bolivian and Peruvian Amazonia, gradually took the lead.[78] In the Lower Amazon other products like jute and hot peppers carried the day, and some of the traditional nut-producing areas showed signs of exhaustion, probably due to the clearing of forest areas adjacent to the castanhais. Political concerns had shifted away from the nut workers and were now focused on the Battle for Rubber, which tried to revive the production of this item for the war effort. In later decades the black peasants and other sectors of Amazonia's society faced new challenges and opportunities, as the federal government promoted the region's development and new forms of labor-based activism spread.[79] Clearly, the rural labor unions that spread through the region in the 1970s, the landless peasant organizations of the 1980s, or the quilombo associations of the 1990s did not operate on virgin soil. Rather, they fed off a series of fragmentary experiences, memories, discourses, and strategies that had been put in practice many decades before.

Conclusion

In his famous 1959 economic history of Brazil, economist Celso Furtado argued that during the colonial period Amazonia was essentially "abandoned by the Portuguese government" and that it "lived exclusively on the forest-extractive economy organized by the Jesuit fathers and based on the exploitation of Indian manpower."[1] He presented the state of Maranhão, renamed as Grão-Pará and Maranhão in 1751, as not only peripheral to the sugar and the mining regions of colonial Brazil but also as "completely isolated" from the relevant economic centers of the Portuguese empire in the Americas. In sum, this region was not only dysfunctional but also devoid of any relevance to the history of Brazil. Having as his main reference the colonial model of the slave-holding economies of Bahia and Minas Gerais, Furtado considered the state of Maranhão to be basically a failed imitation of the social and economic paradigm of plantation society.

More recently, the region has entered the Brazilian and Latin American historiographic canon in the context of the rubber boom. Unfortunately, the new visions of Amazonia have done little to alter traditional perceptions of the region as essentially different from the rest of Brazil in terms of its economic, social, and racial profile. It continues to be seen as a backwater in which nothing important happened, and that only gained prominence during the three or four decades when it supplied North Atlantic markets with latex. Amazonians have been portrayed as historical actors who seem to have done little more than react to external stimuli. Naturally, there are some exceptions, but most studies on Amazonia continue to focus on the outside perspectives of intellectuals, policy makers, entrepreneurs, foreign powers, and the federal government. Few works portray the locals as having any agency.[2]

The People of the River reinserts Afro descendants into Amazonia's history, especially in the periods immediately before and after the rubber boom. In the long century that goes from the eighteenth-century Pombaline reforms to the rubber era, plantation agriculture took off in the region, making African slavery a core component of Amazonia's economic, demographic, social, and cultural landscapes. Between 1800 and 1850, slaves made up around one-third of the state's total population and never less than one-fifth, and by the 1820s a number of municipalities in the Guajarine area counted more than half of their inhabitants as slaves. The Cabanagem revolt of 1835–1840 obviously damaged slaveholding properties, but plantations and ranches employing enslaved Africans and Afro-Brazilians were among the main catalysts of the region's recovery in the following decades. Rubber was only one more product in that period, not the absolute king it would become by the last decades of the century.

Nothing prevented Amazonia from developing a plantation agriculture, and certainly not the region's environment, as we saw in chapter one. Despite the harsh regime of tropical rains and the poor soil in most of the uplands across the Amazon basin, both cacao and sugarcane could be cultivated in the high floodlands, areas that were submerged seasonally for brief periods. Local planters took advantage of the compatible relationship between these two crops and other staples, used the energy generated by the river's tides, and relied on the tide to fertilize their fields. By successfully adapting plantation agriculture to the environmental conditions of the region, then, Amazon planters proved that they could successfully produce export crops, facilitating the expansion of the regional economy in the post-Cabanagem decades. Local labor relations required no modifications to incorporate African slavery either, because Indian slavery was sadly common in Amazonia before captives from the equatorial Atlantic hit its shores. Since the colonial period, different forms of coerced labor were an intrinsic feature of interactions between Western colonists, the state, and indigenous groups. Despite Indian enslavement being almost completely prohibited in 1680, by 1750 Amazonian Indians were still employed making sugar; planting cacao, coffee, and rice; grazing cattle; tapping rubber; collecting forest spices; building fortresses; guiding military forces in the forest; fighting foreign armies; or tending to food gardens—and all the while were usually treated as "virtual slaves."[3] When enslaved Africans arrived in the Amazon basin, then, racial hierarchies and the conduct of forced labor were both well established from generations ago. Had the independence of Brazil and the Cabanagem revolt not severed Belém from the Atlantic slave trade during the period when it peaked, between 1800 and the 1810s, Amazonia could have ended up being a "slave society" relying on captive African laborers, instead of simply a "society with slaves."[4]

In 1840, as the Cabanagem revolt finally ended in the cities and in most rural areas, rubber was already among Pará's exports, but from that year until the mid-1860s or 1870, when rubber became the indisputable king, plantation crops produced by enslaved Africans represented a fundamental source of revenue for the state's economy. Cacao, rice, sugar, and even cotton and coffee made it onto the state's export rubrics during those decades--a list of exports that reads almost like a catalog of plantation items in the history of the Atlantic. As we saw in chapter 1, it is literally impossible to understand the economic recovery that took place between 1840 and 1870 without considering the plantation crops that enslaved Africans produced.

When latex finally surpassed the value of all other exports by the late 1860s, slave production was redirected to the booming internal markets, which were growing at a very fast pace as the state's population swelled. For those toiling in chains, the changes induced by the rise of rubber were a mixed blessing. On one hand, the enslaved could erect thicker parallel economies. As illustrated by the case of Larry, thanks to the commercial bonanza brought by the hevea's milk during the 1870s and 1880s there were more opportunities, both legal and illegal, to engage in the trade in forest items and therefore more transportation, extractive, and guiding activities available—and probably more cash payments as well. On the other hand, by adapting successfully to the changes brought by rubber, slaveholding planters were able to resist the process of abolition until the late 1870s—and in some cases until 1888. Scrambling for any available labor, Paraense slaveholders relied on the plasticity of Brazilian slavery to retain large slave crews, thus maintaining a positive balance of slave arrivals in the state during the last decades before emancipation— something that no other Brazilian province out of the southeast achieved in this time period.

For the most part, the core areas of rubber extraction between the 1870s and 1913 did not overlap with the plantation areas. Neither the municipalities of Breves and Anajás, west of the island of Marajó, nor the valleys of the Xingu and the Tapajós, farther upriver from the Amazon's main course, contained large numbers of enslaved Africans. But despite this lack of overlap, the rubber boom did affect the historical trajectory of Amazonian blacks. Restrictions on the privatization of Brazil nut groves were removed by law in 1910, which paved the way for their acquisition by a series of commercial houses then and during the next decade in the Lower Amazon and the lower Tocantins River area. The collapse of rubber exports in the 1910s only accelerated the search for alternative extractive products—and Brazil nuts figured prominently among them.

In the post-rubber decade of the 1920s, the black peasants of the Lower Amazon and other areas participated in numerous protests and revolts against the demarcations of Brazil nut groves. Proclaiming customary norms embedded in the natural landscape, resorting to traditional bonds with political patrons, and eventually engaging in modern forms of mass protest, they met with local authorities and police forces, attacked foreign merchants, pilfered Brazil nuts, and addressed the state government through representatives like Palma Muniz throughout the decade. In doing so, they joined the disgruntled army officers and the young liberal politicians who in 1924 revolted in Óbidos and in the neighboring state of Amazonas against the rampant corruption and social problems of Brazil's First Republic. Overall, the protests of the Lower Amazon black peasants contributed to the climate of dissatisfaction and instability that ultimately led to the ascent to power of national president Getúlio Vargas in 1930 in Rio de Janeiro, and of federal interventor Joaquim de Magalhães Cardoso Barata in Belém. In sum, while Africans and Afro-descendants had originally arrived in Amazonia as slaves, they did not conform to that role. Instead, they shaped the region's history by building autonomous communities, developing a peasant project, and having an impact on local, state, and even national politics. Natural landscapes, as we see next, played a pivotal role in that process.

Identity: What's Nature Got to Do with It?

The People of the River argues that natural landscapes represented a vehicle for the genesis and the evolution of an Afro-descendant identity among Pará's black peasants. The process of developing that identity gained force in the last decades of slavery, when the ties of cooperation and support that the enslaved had built among them generated a shared culture and some common aspirations.[5] As I aimed to show here, the material and symbolic dimensions of natural landscapes were pivotal to the parallel economies and the bonds of community that the enslaved erected in senzalas and maroon communities. Tying the famous manioc grounds cultivated on the uplands with the collection of birds, small mammals, reptiles, amphibians, insects, mollusks, nuts, wild fruits, medicinal plants, fibers, oils, rubber, timber, exotic products, and many other forest items, black Amazonians slowly but surely learned how to carve a life as independent farmers, even before freedom officially arrived. Developing a complex body of knowledge about the forest was a potential act of subversion: by doing so an individual could achieve a specialized profession, improve his or her living conditions, and form a stable family. Because forest skills were "currencies useful in ranking individuals within the local social structure," "the physical environment of the plantation" became "part of the [slaves'] elaborate geography of kinship."[6] To the point, I would add, of reverting the role of "mechanical energy" providers that the slaveowners had planned for the African bondsmen.[7]

There were many paths toward environmental creolization. Amazonia is a vast tropical forest with different subregional variations, such as rainforests, savannahs, mangroves, and flooded forests—which made experiences brought from equatorial Africa useful in the New World. Amazonia's portfolio farming system—the seasonal combination of multiple agricultural and collection activities—was not new to the Africans who came in chains to its shores. If they could form a family, the enslaved would also replicate the household labor system that they practiced in Congo or Angola. Moreover, as the slave trade tied West Central Africa and Amazonia closer together in the late eighteenth and early nineteenth centuries, the rural livelihood of farmers on both Atlantic shores grew even more similar. Manioc cultivation and processing, for example, became key to food production on both sides of the ocean. Thus, although West Central Africans were newcomers to Amazonia, their knowledge of farming and forest collection was not.[8] In turn Indians could also teach arriving Africans the secrets of the forest. Because the Indians were supposed to be more productive when working alongside enslaved Africans, many planters favored the cohabitation of both labor forces. However, the sharing of living spaces, labor routines, knowledge, opinions, projects, traditions, and even households would actually empower the African captives in their drive for freedom and autonomy, as they enhanced their knowledge of natural landscapes thanks to interaction with the natives.

From the standpoint of both slaves and maroons, the growing network of itinerant merchants known as regatões was an added stimulus to the construction of a parallel economy—and one that would continue to grow during the second half of the century. Networks of clandestine relations with commercial agents represented a source of food, manufactures, weapons, and cash for the enslaved all over the Americas, but in Amazonia the vigor of these networks and the great extent of the watercourses they used as roads made them a formidable resource in the struggle for autonomy and freedom. Instead of an "internal economy" of slavery, the slaves' productive and commercial endeavors were just one facet of a vast, far-reaching, and sophisticated economic system *parallel* to that of the slaveowners.

The process of environmental creolization also took place in the realm of culture. Consider the narrative of the Big Snake employed by maroons in the Trombetas and Cuminá River areas to narrate their history. The story has Bantu African elements for sure, but the influence of Kaxúyana and broader Amazonian mythology about giant aquatic snakes is beyond doubt. In a process of cross-fertilization between local traditions and imported ones, Lower Amazon maroons embedded their history in local landscapes, appropriating them to carry out a cultural, and almost political, process of generating a sense of belonging through nature. Long after the first generation of maroon pioneers was gone, "the social memory of slavery [and of marronage] evolved and was maintained through narratives and [through] the retrospective memory shaped by landscape and built structures."[9]

Differences between groups of maroons and plantation slaves lost importance once freedom came: neither the shackles of slavery nor the threat of capture constrained their peasant projects anymore—although, of course, poverty and a lack of formal access to political power did. After 1888, both ex-slaves and ex-maroons fomed black rural communities characterized by the existence of thick kinship networks, the practice of household-based subsistence agriculture, and a remarkable emphasis on autonomy; that is, on the ability to freely engage in trade and in wage work.[10] Regardless of their origins, in other words, they ended up forming a black peasantry. Nonetheless, these similarities did not transform black peasants into a group defined by essentialist and inmutable characteristics. While they shared many characteristics with other sectors of the broader Brazilian peasantry, internally black peasants also had fault lines based on ethnic origin, race, gender, income, status, and religion.[11]

As they started a life in freedom and experienced conflicts with merchants and landowners, the inhabitants of communities like Pacoval or Cacau established political alliances and shared discourses with other segments of the Brazilian citizenry. I discussed, for instance, the existence of a traditional sphere of relationships with members of local elites, which in some cases permitted a precarious but real degree of influence over local institutions. Consider the ties between the community of Pacoval and local politician Fulgêncio Simões, for example. Those ties brought official state authorities to Pacoval and nearly produced an official collective land deed for its inhabitants. *Mateiros, prepostos, fiscais,* and other privileged agents of Brazil nut traders in the Trombetas and Cuminá River areas turned out to be more of an obstacle for black aspirations, but they shaped their political strategies too. The inhabitants of the community of Cacau, who went in person to visit the new owner of their lands when their living arrangements changed, were trying to preserve similar relationships with traditional landowners, who eventually became mythical patrons. Despite the lack of access to formal political participation through the vote, then, black peasants used patronage relations to advance their personal and collective agendas.

Popular sectors of Amazonian society also provided useful discursive and juridical categories to the black peasants. The idea that customary land

tenure was conducive to the recognition of property rights has always circu-
lated among rural Brazilians, including the black rural groups discussed here.
The wave of nativist protests occurring all over the country in the 1920s also
represented a new opportunity to solidify links locally and regionally with
commercial houses that loathed the newly arrived European merchants, as
shown by the 1921 Pacoval protests. In response to the rapid ascent of Italian
and Portuguese immigrants, Brazilian-born merchants and mocambeiro nut
extractors found common ground to formulate nativist claims based on the
recognition of their shared Brazilianness.

But as this book shows, the repertoire of ideas, discourses, and strategies
extracted from the natural world remained key to the black peasants in the
post-emancipation era. The Trombetas River, for example, was not just the
location where the maroons became free peasants—it was an active participant
in it. The river constituted a strange and deadly place initially. Later it became
a weapon against slaveholders and a living embodiment of syncretic deities.
By the time the peasants relocated below the waterfalls in the years around abolition, the river was a valuable economic resource. The flogging trunk of the
Campina sugar plantation, in Vigia, silently articulated the changes in the landscape that took place over time. While it originally functioned as a linchpin
for slave work at the plantation, two generations later this natural site had become a symbol of the ties linking black peasants to a shared past in slavery
expressed through songs and popular narratives. "The natural environment
was not merely a scene in which action took place," but was instead directly
"entangled with action."[12]

During the Brazilian First Republic there were a number of serious conflicts
over land and labor, as waves of capitalist modernization modified relations
of production all over the country. When Afro-Amazonians defended their
rights as Brazilians vis-à-vis state authorities, they relied on narratives of belonging and invoked past struggles against slavery through tropes of nature.
Thus, the black peasants claimed to be "citizens of Tauapará," "the people of
the Curuá," or "sons of the Trombetas" River. They also demanded from state
representatives that access to forest resources remain "free," using a concept
with powerful resonances. In doing so, they merged nature with the rights of
citizenship, thus forming political agendas with an ethno-ecological conscious-
ness. The natural world was, in sum, a core component of the discursive strategies adopted by black peasants during the First Republic's struggles for
citizenship. Those political strategies did not happen only in a natural landscape; they happened *through that landscape*.

Roasted Duck or the Novelty of Black Peasants' Traditions

When they were advanced in the First Republic, the black peasants' claims over land and citizenship were interpreted as the result of their being manipulated by "adventurers" seeking "the greatest gain with the least investment, paying the local workers miserably." Black peasants were seen as unable to formulate their own political agenda and were perceived to be members of backward, poor rural populations tainted by a mixture of poverty, racial inferiority, and rural backwardness. Intellectuals like Aluísio de Azevedo or Nina Rodrigues espoused ideas about the alleged inferiority of Africans caused by a combination of racial and cultural factors. Other prominent intellectuals of the First Republic, such as Oliveira Vianna and Euclides da Cunha, also highlighted the "deplorable mental situation" of the "sub-races of the Brazilian backlands," "probably . . . destined to disappear soon," given the influx of European immigrants to Brazil in the late 1800s. The atavistic, irrational, and archaic non-white rural Brazilians were, after all, the protagonists of the millenary revolts of Canudos, Contestado, and Juazeiro.[13]

By the mid-twentieth century, the emphasis on class as a category of analysis in rural conflicts led black rural communities to be characterized as simply peasants. Adopting a Marxist perspective, Brazilian scholars in the mid-1900s conceived rural populations as social actors whose conflicts and ideas were mainly—and sometimes exclusively—determined by their structural relation to the means of production.[14] Moreover, heated disputes between agribusiness and landless peasants in the middle decades of the century gave weight to the idea that access to landownership was the top unresolved social problem in the country since the colonial period.[15] While the ethnic background of Brazilian peasants mattered in the study of European immigrants and indigenous peoples, with a few notable exceptions scholars did not pay much attention to the legacies of slavery and African culture for the living conditions of twentieth-century rural Brazilians.

The lack of salience of racial and ethnic elements in these analyses makes sense if we consider the high degree of racial mixture that characterizes rural populations in Brazil. Traditionally, they have been seen as a mixture of blacks, Indians, and whites, defying easy identification with any single racial category. During the mid- to late twentieth century, black peasants also participated in the same political and class-based organizations as the rest of the peasantry, sharing nationalist and occasionally revolutionary ideas. Rural labor unions from all over Brazil always boasted militants of all colors—much like urban ones did. The nationalist, class-oriented labor organizations that have predominated for most of the twentieth century, such as the Ligas Camponesas or the more modern Landless Workers' Movement (MST), have traditionally embraced—and to a good extent practiced—the vision of a national community where race should not matter.[16]

More recently, some scholars have emphasized how the passing of Article 68 in the 1988 Constitution has led present-day peasant communities to craft a new collective identity aligned with institutional changes and black political activism. By giving land to quilombo descendants and by recognizing them as the subject of special ethnic rights, Article 68 inspired a process of "ethno-racial identity formation," instituting "a new and ethnically differentiated political subject" in the Brazilian countryside.[17] It led to a "moment of emergence of a collective identity," the "genesis of new leaders," the adoption of new political "grammars," the development of new ties with governmental agencies, participation in international advocacy networks, and eventually changes in the collective memories of a number of communities.[18] In some cases, the anthropologists who study such processes have also recognized that those new identities were influenced by preexisting forms of ethno-political representation. For example, in her analysis of how the law molded local identities in the community of Mocambo in northeastern Brazil, Jan Hoffman French observes that traditional identities did not disappear when new ones emerged, but rather continue to matter in what she calls an "axis of representational durability."[19]

There is little doubt that the availability of new resources and opportunities at the national level has been a catalyst for an entire wave of black activism in the so-called (due to the wording of Article 68) quilombo descendant communities. During the 1990s, for example, to fight the encroachments of mining and agribusiness companies, the Lower Amazon black communities resorted to the new quilombo clause in their efforts to obtain formal deed to their lands. The community of Boa Vista, in the Trombetas River, pioneered this process in 1992. Located near the mining complex of Porto Trombetas, which belonged to the Mineração Rio do Norte (MRN) consortium of Brazilian and American companies, Boa Vista suffered the rapid diminution of its traditional hunting, fishing, and collecting areas, and a number of its inhabitants became wageworkers for MRN in poor conditions. With the help of Verbite liberation theology missionaries, rural labor union militants, and urban black activists from Belém's CEDENPA, Boa Vista started an application for recognition and land titling as a *remanescente de quilombo*. In 1995, during the many government-backed celebrations of the 300th anniversary of maroon leader Zumbi dos Palmares's death in Brasília, Boa Vista was among the first three communities in Brazil to receive a collective land deed and be recognized as a remanescente de quilombo.[20] While initially Article 68 seemed to be addressed to maroon descendant communities only, over time it was deployed to eventually target all black communities in Brazil.

Although Article 68 sparked a process of black ethnic reconfiguration, the emphasis on the novelty of such identities has inadvertently obscured the vitality of black political traditions. In the Boa Vista community, for example, neither its conflicts with MRN nor the community's political mobilization began in 1988, when Article 68 was passed. Rather, during the 1970s and 1980s the community had allied with other actors in the region, such as the rural labor unions and the Verbite missionaries, to fight against the land grabs of land speculators and to demand protection from the authorities. The conflicts of the 1990s overlapped in the memory of the mocambeiros with those of the 1970s and 1980s, and those in turn interacted with older ones from earlier eras.

One final anecdote illustrates my point. As I was conducting interviews in 2009 in the maroon-descendant community of Pacoval in Alenquer, something troubled me. In my conversations with Dona Cruzinha, one of its leaders, I was encouraging her to share the oral traditions about the 1921 conflict described in this book. However, Dona Cruzinha kept referring to two other conflicts: one in the 1950s with Portuguese merchant Joaquim Tavares de Sousa and another one from the 1970s involving a rebellious nut extractor named *Pato Assado,* or Roasted Duck, who "fought for the freedom of the castan-hais." During the interviews I conducted in those weeks in Pacoval I kept trying to disentangle the 1921 conflict from the one in the 1950s, and those two from the Pato Assado one. After several of my informants laughed as soon as I brought up Pato Assado or assigned him to different time periods, I gave up trying to disentangle those conflicts. Although written documents proved they had occurred decades apart, they had merged into a single genealogy of ethno-racial politics. The Pacovalenses had bound together different struggles over land and community through their fierce defense of natural resources and through their use of natural metaphors to speak of politics.

In sum, as *The People of the River* shows, the political actions of the 1990s were just a new iteration of a much older tradition of black peasant politics dating at least from the era of slavery. This applies not only to the those in the Trombetas River area, but also to a host of black peasant communities throughout Pará and the rest of Brazil who descend from plantation slaves, maroons, and freed people. The "political opportunity"[21] that opened up in 1988 through Article 68, in other words, did not fall on deaf ears. While discourses defending the rights of citizenship for Afro-descendants were revamped to accommodate to the new era that Article 68 inaugurated, they continued to refer to environmental tropes that have been used in previous key moments. Black peasants continue to assert their rights as Brazilians through multiple dialogues, but their voice has always found in nature a vehicle to maintain a singular identity along the way.

Notes

Introduction

1. CENTUR, Muniz, *Castanhaes de Alemquer*, 45, 16, 46–47.

2. "Anexo: Associação das Comunidades Remanescentes de Quilombos do Município de Oriximiná," Acevedo Marin and Castro, *Negros do Trombetas*, 263.

3. "Aos remanescentes das comunidades dos quilombos que estejam ocupando suas terras, é reconhecida a propriedade definitiva, devendo o Estado emitir-lhes títulos respectivos." Available at www.cpisp.org.br/htm/leis/index.html. Accessed May 11, 2011. Articles in the Transitory Dispositions are meant to disappear once they have served their purpose.

4. A good summary of the titles for the Trombetas communities is at Andrade, *Quilombola Lands*, 13–14. Their official name of the communities is *remanescentes de quilombo* or maroon descendants, but I call them simply black rural communities because only some are made up of descendants of strictly defined maroons. Sistema de Monitoramento do

Programa Brasil Quilombola, "Quadro geral de Comunidades Remanescentes de Quilombos (CRQs)"; Instituto Nacional de Colonização e Reforma Agrária, "Relação de processos," 6; Instituto Nacional de Colonização e Reforma Agrária, "Perguntas e Respostas," 10.

5. De la Torre, "Are They Really Quilombos?," 102–5; French, *Legalizing Identities*, 6.

6. This is the phrasing of Decree 4,887 from 2003, regulating the application of Article 68. Malcher and Ataide, *Territórios Quilombolas*, 3:44; SEPPIR, *Programa Brasil Quilombola*, 6.

7. Linhares, "Kilombos of Brazil," 831; Treccani, *Terras de Quilombo*, 121; Véran, "Quilombos and Land Rights," 22–23. A few examples of their sustained presence in the media include Abrams, *Quilombo Country*; Bowater, "Brazil's Quilombos"; Brooke, "Brazil Seeks to Return Ancestral Lands"; Coutinho, Paulin, and Medeiros, "A farra"; Fairbanks, "The Global Face"; Garcia-Navarro, "For Descendants of Brazil's Slaves"; Gutiérrez, "Negros con título de propiedad"; Mann and Hecht, "Brazil's Maroon People"; Marull, "Tres millones de descendientes"; Planas, "Brazil's 'Quilombo' Movement"; SEPPIR, *Programa Brasil Quilombola*. "Between 1995 and 1997, seventy-three new books, theses and dissertations, monographs, and articles appeared on the theme" of quilombos, according to Slater, *Entangled Edens*, 268.

8. Barbara, "In Denial over Racism"; Cottrol, *The Long, Lingering Shadow*, 243; Telles, *Race in Another America*, 60.

9. French, *Legalizing Identities*, 3.

10. Quotations from Kenny, "The Contours of Quilombola Identity," 142, 154–55; Carvalho, "O Quilombo da 'Família Silva,'" 37, 42–43; see also Alonso, "O 'movimento' pela identidade," 27–28; Mattos, "Políticas de reparação e identidade coletiva no meio rural," 181; Véran, "Rio das Rãs," 316, 318.

11. Candido, *Os parceiros do Rio Bonito*; Hutchinson, *Village and Plantation Life*; Fry and Vogt, *Cafundó*. There was an avalanche of monographs on modern quilombos. For overviews of these studies, see O'Dwyer, *Quilombos: Identidade Étnica e Territorialidade*; Arruti, "Comunidades remanescentes de quilombos"; Associação Brasileira de Antropologia, *Territórios Quilombolas*; Carvalho, *O Quilombo do Rio das Rãs*; Gusmão, "Herança Quilombola"; Dutra, *Direitos Quilombolas*; Price et al., "Dossiê Remanescentes de Quilombos"; Schmitt, Turatti, and De Carvalho, "A Atualização do Conceito de Quilombo: Identidade e Território nas Definições Teóricas"; SEPPIR, *Programa Brasil Quilombola*; Treccani, *Terras de Quilombo*.

12. The word "maroon" is used both for the community and for the individuals living in it. Classic studies include Agorsah, *Maroon Heritage*; Bilby, *True-Born Maroons*; Heuman, *Out of the House of Bondage*; Izard, *Orejanos, Cimarrones, y Arrochelados*; La Rosa Corzo, *Runaway Slave Settlements in Cuba: Resistance and Repression*; Landers, "Gracia Real de Santa Teresa de Mose"; Laviña, "Comunidades Afroamericanas"; Price, *First-Time*, 1983; Price, *Maroon Societies: Rebel Slave Communities in the Americas*; Reis and Gomes, *Liberdade por um Fio*; Thompson, *Flight to Freedom*.

13. Recent analyses of marronage have emphasized these complex relationships: Barcia, *Seeds of Insurrection*, chap. 3; Beatty-Medina, "Between the Cross and the Sword"; Bryant, *Rivers of Gold*; Gomes and Machado, "Migraciones, desplazamientos y campesinos negros"; Landers, "Cimarrón and Citizen"; Lentz, "Black Belizeans and Fugitive Mayas"; Lockley, *Maroon Communities in South Carolina*; Miki, "Fleeing into Slavery"; Rupert, "Marronage, Manumission."

14. Quotations from Greider and Garkovich, "Landscapes," 1; Rogers, *The Deepest Wounds*, 6; Baker, "Introduction," 9; Tilley, "Introduction," 18; see also Blomley, "Landscapes of Property," 574–76; Heath and Lee, "Memory, Race, and Place," 1353; Raffles, *In Amazonia*, 8, 54–55, 74.

15. Quotation from Machado, *O plano e o pânico*, 22. See also Gomes, "Slavery, Black Peasants," 745; Mattos, *Das cores do silêncio*, 52, 64, 141, 153; Slenes, *Na Senzala, Uma Flor*, 248.

16. Quotation from Gomes and Machado, "Migraciones, desplazamientos y campesinos negros," 31. See Fraga Filho, *Encruzilhadas da liberdade*, 212, 223, 253; Gomes, "Roceiros, Mocambeiros," 165; Gomes, "Slavery, Black Peasants," 745; Guimarães, *Múltiplos Viveres*, 286; Miki, "Fleeing into Slavery," 528; Naro, *A Slave's Place, A Master's World*, 159; Rodrigues, "Serra dos Pretos," 173. For analogous processes in other regions, see Penningroth, "The Claims of Slaves and Ex-Slaves"; Scott and Zeuske, "Demandas de Propiedad y Ciudadanía."

17. Mattos, *Das cores do silêncio*, 52; Gomes, "Slavery, Black Peasants," 745.

18. Quotations from Gomes, "Slavery, Black Peasants," 745; Mahony, "Creativity under Constraint," 657; Slenes, *Na Senzala, Uma Flor*, 208. On the internal economy and its implications, see Barickman, "A Bit of Land"; Berlin and Morgan, *The Slaves' Economy*; Cardoso, "The Peasant Breach in the Slave System"; Chalhoub, *Visões da liberdade*, 59, 68, 80; Fraga Filho, *Encruzilhadas da liberdade*, 192; Gomes, "Slavery, Black Peasants," 745; Miki, "Fleeing into Slavery," 506–11; Slenes, *Na Senzala, Uma Flor*, 150, 197–208. On the rich meanings of family among the enslaved, see Florentino and Góes, *A paz das senzalas*, 45, 124; Mattos, *Das cores do silêncio*, 141; Miki, "Fleeing into Slavery," 525–27; Rios and Mattos, *Memórias do cativeiro*, 299; Slenes, *Na Senzala, Uma Flor*, 208; Sweet, "Defying Social Death."

19. On this, see chapter 3, notes 24–28.

20. For example, see AMO, Livro de Sessões do Conselho Municipal, 1840–1858, July 8, 1850; Magalhães, *Relatório dos negocios*, 9; Gomes, *Nas Terras do Cabo Norte*, 396, 399.

21. Comprehensive and enlightening discussions of the term can be found at Assunção, *Capoeira*, 32–46; Parés, "O Processo de Crioulização," 87–97.

22. Quotations from Carney, "Landscapes of Technology Transfer," 5; on the Lowcountry, from Brown, *African-Atlantic Cultures*, 2; on "creole ecologies," from McNeill, *Mosquito Empires*, 30; and on "highway and sanctuary," from Stewart, "Slavery and the Origins," 15. See also Carney, *Black Rice*; Carney, "'With Grains in Her Hair'"; Carney and Rosomoff, *In the Shadow of Slavery*; Hawthorne, *From Africa to Brazil*; Hawthorne, "From 'Black Rice' to 'Brown'"; Silver, *A New Face on the Countryside*, 106, 135, 144.

23. Dean, *With Broadax and Firebrand*, chaps. 3–8; Funes Monzote, *From Rainforest to Cane Field in Cuba*; McNeill, "The Ecological Atlantic"; Merchant, *American Environmental History*, chap. 3; Rogers, *The Deepest Wounds*, 32–44; Silver, *A New Face on the Countryside*, chaps. 5–6; Soluri, *Banana Cultures*; Stewart, *What Nature Suffers to Groe*, chap. 3; Tucker, *Insatiable Appetite*, 15–119; Watts, *The West Indies*, 393–517.

24. Chalhoub, *Visões da liberdade*, 26; Gomes and Cunha, "Introdução," 9–11.

25. Dean, *With Broadax and Firebrand*, 57.

26. Querino, "O colono preto," 156–57; Freyre, *The Masters and the Slaves*, 70; Fernandes, *The Negro in Brazilian Society*, 28, 67.

27. Parés, "O Processo de Crioulização," 131; Schwartz, *Sugar Plantations*, 251; Florentino and Góes, *A paz das senzalas*, 35, 175; Reis, *Slave Rebellion in Brazil*, 121–23; Sweet,

Recreating Africa, 34–50; Faria, "Identidade e comunidade escrava," 145; Mattos, *Das cores do silêncio*, 141.

28. Quotations from Fraga Filho, *Encruzilhadas da liberdade*, 290; Moreira and Hébette, "Metamorfoses de um campesinato," 193; Rios, "My Mother Was a Slave," 116; Paoliello, "Condição camponesa," 246.

29. Quotations from Gomes, "Roceiros, *Mocambeiros*," 164–65; Faria, "Identidade e comunidade escrava," 145. See also Boyer, "Misnaming Social Conflict," 527, 530; Farfán-Santos, "'Fraudulent' Identities," 124–25, 128–30.

30. Greider and Garkovich, "Landscapes," 8; Giesen, "The Truth about the Boll Weevil," 684.

31. Peck, "The Nature of Labor," 230; Rogers, *The Deepest Wounds*, 68; Heath and Lee, "Memory, Race, and Place," 1360.

32. Veríssimo, "Ethnographia," 135; Furtado, *The Economic Growth of Brazil*, 73, 96; Prado Jr., *The Colonial Background*, 244; MacLachlan, "African Slave Trade," 116; Fausto, *Brasil, de Colonia a Democracia*, 50. There were exceptions too: economist Roberto Simonsen and anthropologist Charles Wagley, for example, recognized and studied the black presence in the region: Simonsen, *História econômica*, 34:440–47; Wagley, *Amazon Town*, 129–42.

33. Chambouleyron, "Escravos do Atlântico Equatorial," 101–2; query slaves disembarked until 1702 in Amazonia, "Voyages: The Trans-Atlantic Slave Trade Database," www.slavevoyages.org/voyage/search. Accessed January 2, 2011.

34. Query "1680–1702" and "Brazil" as principal place of slave landing. "Voyages: The Trans-Atlantic Slave Trade Database," Accessed April 14, 2014.

35. De Souza Junior, *Tramas do Cotidiano*, 89–90, 75. See also Costa, "Lugar e significado"; Maxwell, *Pombal*; Rodrigues, "Para o Socego e Tranquilidade Publica."

36. Dias, *Fomento e mercantilismo*, 1:211–23.

37. "Introducção Secretissima com que Sua Magestade manda passar á Capital de Belém do Grãa-Pará o Governador e Capitão-General João Pereira Caldas," Rei and Conselheiro Manoel José Maria da Costa e Sá, September 2, 1772, in Moraes, *Corographia Historica*, 2:141.

38. Hawthorne, *From Africa to Brazil*, 52–53.

39. Hawthorne, *From Africa to Brazil*, 140, 174, 234.

40. Query slaves disembarked in Pernambuco, 1800–1810, "Voyages: The Trans-Atlantic Slave Trade Database," www.slavevoyages.org/voyage/search. Accessed April 2, 2014.

41. Silva, "The Atlantic Slave Trade," 488; Eltis and Richardson, "A New Assessment," 20.

42. De Souza Junior, *Tramas do Cotidiano*, 316; Harris, *Rebellion on the Amazon*, 139.

43. Eltis and Richardson, "A New Assessment," 19.

Chapter One

1. Belmar, *Voyage aux Provinces Brésiliennes*, 132.

2. Alden, "The Significance of Cacao Production," 131; Furtado, *Formação Econômica do Brasil*, 129–30; Santos, *História econômica da Amazônia*; Weinstein, *The Amazon Rubber Boom*. My own interpretation is indebted to Anderson, "Following Curupira," 68–72; Batista, "Muito Além dos Seringais," 69–70.

3. Alden, "The Significance of Cacao Production," 155; Mahony, "The World Cacao Made," 218; Roller, "Colonial Collecting Expeditions," 435. Recent findings indicate that cacao had been cultivated in the 1700s as well: Chambouleyron, "Cacao, Bark-Clove and Agriculture," 9–10.

4. Alden, "The Significance of Cacao Production," 132; Chambouleyron, "Cacao, Bark-Clove and Agriculture," 10; Harris, *Rebellion on the Amazon*, 135–37.

5. Maw, *Journal of a Passage*, 341; Bates, *The Naturalist on the River Amazons*, 123. See also Brown and Lidstone, *Fifteen Thousand Miles*, 97–98; Harris, *Rebellion on the Amazon*, 135–38.

6. On cacao varieties, see Bondar, *A cultura do cacau na Bahia*, 11; Le Cointe, *L'Amazonie Bresilienne*, 2:126–27; Mahony, "The World Cacao Made," 49; Young, *The Chocolate Tree*, 42–46.

7. Cordeiro, *O Estado do Pará*, 61; Pará, *Relatorio 1867*, 17; Le Cointe, *L'Amazonie Bresilienne*, 2:141–42. Compare to data on rubber exports from Weinstein, *The Amazon Rubber Boom*, 53.

8. Anderson, "Following Curupira," 295–303. On Bahia, see Mahony, "The World Cacao Made," 205.

9. Bates, *The Naturalist on the River Amazons*, 104; Young, *The Chocolate Tree*, 2; Le Cointe, *L'Amazonie Bresilienne*, 2:116. On cacao growing in the shade of taller trees, see Bates, *The Naturalist on the River Amazons*, 139; Bondar, *A cultura do cacau na Bahia*, 6, 26; Le Cointe, *L'Amazonie Bresilienne*, 2:116; Morais, *Apontamentos de viagem*, 227; Smith, *Brazil, the Amazons and the Coast*, 111.

10. On the necessity of abundant organic matter, see Bondar, *A cultura do cacau na Bahia*, 9; Herndon, *Exploration of the Valley of the Amazon*, 298; Le Cointe, *L'Amazonie Bresilienne*, 2:118; Young, *The Chocolate Tree*, 162–63. On cultivation in the high floodlands, see Bondar, *A cultura do cacau na Bahia*, 17; Bates, *The Naturalist on the River Amazons*, 140; Brown and Lidstone, *Fifteen Thousand Miles*, 197; Le Cointe, *L'Amazonie Bresilienne*, 2:116; Morais, *Apontamentos de viagem*, 232; Smith, *Brazil, the Amazons and the Coast*, 94.

11. Bondar, *A cultura do cacau na Bahia*, 30; Le Cointe, *Amazônia Brasileira III*, 88. On the two annual harvests, see Alden, "The Significance of Cacao Production," 115; Herndon, *Exploration of the Valley of the Amazon*, 299; Le Cointe, *L'Amazonie Bresilienne*, 2:124; Smith, *Brazil, the Amazons and the Coast*, 111.

12. Herndon, *Exploration of the Valley of the Amazon*, 298–99; Le Cointe, *L'Amazonie Bresilienne*, 2:124–33; Smith, *Brazil, the Amazons and the Coast*, 111; Young, *The Chocolate Tree*, 74–76. On the cacao "wine," see Herndon, *Exploration of the Valley of the Amazon*, 298; Morais, *Apontamentos de viagem*, 226, 232; Smith, *Brazil, the Amazons and the Coast*, 111.

13. Herndon, *Exploration of the Valley of the Amazon*, 299; Bates, *The Naturalist on the River Amazons*, 139.

14. Pará, *Relatorio . . . 1875*, 61.

15. APEP-Orf, Joaquinna de Moraes Sarmento PMI, 1853; APEP-Orf, Carlos Vicente Faro and Widower PMI, 1853; APEP-Orf, Catharina do Nascimento Silva PMI, 1855; APEP-Orf, Joanna dos Santos PMI, 1858; APEP-Orf, Francisco da Silva PMI, 1859; APEP-Orf, João Maciel da Silva PMI, 1859; APEP-Orf, Manoel Basilio de Carvalho PMI, 1859; APEP-Orf, Thereza de Jesus Negrão PMI, 1859; APEP-Orf, Antonia Maria das Neves PMI, 1860; APEP-Orf, Anna Raimunda Ferreira Pastana PMI, 1864; APEP-Orf, Maria

Victoria dos Santos PMI, 1865; 5A-CR, Antonio José de Souza PMI, 1870; CMA, Manoel José da Silva PMI, 1873; CMA, José Ferreira Bello PMI, 1878.

16. For the first type, see APEP-Orf, Joaquinna de Moraes Sarmento PMI, 1853. For the second, see APEP-Orf, Carlos Vicente Faro and Widower PMI, 1853; APEP-Orf, Francisco da Silva PMI, 1859; APEP-Orf, Thereza de Jesus Negrão PMI, 1859, for example.

17. APEP-Orf, Thereza de Jesus Negrão PMI, 1859; APEP-Orf, Carlos Vicente Faro and Widower PMI, 1853; APEP-Orf, Antonia Maria das Neves PMI, 1860; APEP-Orf, Francisco da Silva PMI, 1859; APEP-Orf, Joaquinna de Moraes Sarmento PMI, 1853; APEP-Orf, Anna Raimunda Ferreira Pastana PMI, 1864.

18. APEP-Orf, José Antônio Lourinho PMI, 1856; APEP-Orf, Anna Raymunda Lobato and Widow PMI, 1856; APEP-Orf, Maria Ritta Corrêa de Miranda PMI; 1857; APEP-Orf, Antônio José de Souza Lima Nunes PMI, 1864; CMA, Antônia Valente das Neves PMI, 1870; CMA, Leonardo José da Costa PMI, 1872; CMA, João Manoel da Silva PMI, 1874; APEP-1V, Maria Felippa de Morais PMI, 1875; APEP-2V, Autos de Partilha Manoel Theodoro de Souza Pinheiro, 1876; APEP-2V, Bartholomeu José de Vilhena PMI, 1877.

19. APEP-Orf, Antonio José Lourinho PMI, 1856; APEP-Orf, Maria Ritta Corrêa de Miranda PMI, 1857.

20. APEP-Orf, Anna Raymunda Lobato and Widow, 1856.

21. Graham, *Caetana Says No*, 9–10, 41; Weinstein, *The Amazon Rubber Boom*, 76, 138, 234.

22. APEP-Orf, Antonio Carlos de Oliveira Bello PMI, 1852; APEP-Orf, Isabel Maria de Moraes PMI, 1857; APEP-Orf, Marcellina Josefa Ferreira PMI, 1859; APEP-Orf, Pedro Baptista de Souza Aranha PMI, 1859; APEP-Orf, Manoel da Cunha Vidinha PMI, 1859; APEP-Orf, Manoel Raimundo dos Santos Quaresma PMI, 1860; APEP-Orf, Antonio Joaquim Pinheiro PMI, 1864; APEP-Orf, Anna Aracema de Jesus PMI, 1864; APEP-Orf, Antonio Ferreira Vaz PMI, 1864; APEP-1V, Felizardo dos Santos Quaresma PMI, 1878; CMA, Antonio Francisco Correa Caripuna PMI, 1877; CMA, Federico Carlos Rhossard PMI, 1878.

23. Anderson, "Sugarcane on the Floodplain," 30–33.

24. Brown and Lidstone, *Fifteen Thousand Miles*, 97–98; Anderson, "Sugarcane on the Floodplain," 34. See also Edwards, *A Voyage up the River Amazon*, 71; Morais, *Apontamentos de viagem*, 232. The combination of cacao and sugarcane is also found in cacao-producing areas of Venezuela: Piñero, *The Town of San Felipe*, 5.

25. Anderson, "Sugarcane on the Floodplain," 37. For the cacao harvests, see Herndon, *Exploration of the Valley of the Amazon*, 299; Le Cointe, *L'Amazonie Brésilienne*, 2:124; Smith, *Brazil, the Amazons and the Coast*, 111. The two harvest periods in Amazonia were very similar to those in Bahia: Mahony, "The World Cacao Made," 320.

26. The sugarcane harvest could be done from May to August as well, but it could be pushed back or forward several months once the cane had been growing for one year.

27. Dean, *Brazil and the Struggle for Rubber*, 54–58.

28. Apparently, large concentrations of the palm were the equivalent to an "all you can eat" buffet for the kissing bug or *Triatoma infestans*, known as *barbeiro* in Brazil and *vinchuca* in Spanish-speaking countries. This insect caused outbreaks of Chagas disease in Mazagão in 1996, in Igarapé-Miri in 2002, in Santarém in 2006, and in Belém in 2007. Vasconcelos et al.,

"Práticas de Colheita e Manuseio do Açaí," 19; Nobrega et al., "Oral Transmission of Chagas Disease"; Pereira et al., "Transmission of Chagas Disease," 75–76.

29. For examples of separated cane fields, see APEP-Orf, Antonio Carlos de Oliveira Bello PMI, 1852; APEP-Orf, Isabel Maria de Moraes PMI, 1857; APEP-Orf, Anna Joaquina Rosa dos Santos Smith PMI, 1857; APEP-Orf, Francisco Ezequiel Sarmento PMI, 1859; APEP-Orf, Marcellina Josefa Ferreira PMI, 1859; APEP-Orf, Raimundo Nonato Roberto Maues PMI, 1860; APEP-Orf, Antonio Joaquim Pinheiro PMI, 1864; APEP-Orf, Anna Aracema de Jesus dos Passos PMI, 1864; CMA, Antonio Francisco Correa Caripuna PMI, 1877; Anderson, "Sugarcane on the Floodplain," 37.

30. On tide mills, see Azevedo, *Arquitetura do açúcar*, 37; Minchinton, "Early Tide Mills"; Nicholson, *The Operative Mechanic*, 94–128, 104–5; Skinner, "Tide Mills of Easton, MD." On the two overlapping tidal cycles, see Anderson, "Sugarcane on the Floodplain," 31.

31. Marques, "Modelo da Agroindústria Canavieira Colonial," 34; Marques and Anderson, "Engenhos movidos a maré," 295–301.

32. For buildings and activities that were usually part of sugar mills, see Armstrong and Reitz, *The Old Village and the Great House*, 87–113; Azevedo, *Arquitetura do açúcar*, 64–78, 106, 155–58; Schwartz, *Slaves, Peasants, and Rebels*, 116–25. They often included a chapel too: APEP-Orf, Maria Thereza de Moraes PMI, 1858; APEP-Orf, Anna Raimunda Ferreira de Pastana PMI, 1864; APEP-Orf, Anna Aracema de Jesus Dos Passos PMI, 1864; APEP-Orf, Angela Maria de Goes PMI, 1865; APEP-IM, Francisca de Monteiro de Noronha PMI, 1865; CMA, João de Figueiredo Muniz PMI, 1871; Maw, *Journal of a Passage*, 375.

33. Amaral, *Memorias*, 1:231; Marques, "Modelo da Agroindústria Canavieira Colonial," 81, 107, 119.

34. On Patroni, see Bezerra Neto, "Por todos os meios," 112–23; Harris, *Rebellion on the Amazon*, 182–85. On Zagalo, see ibid., 128. And on Batista Campos, see Freitas, *La Revolución de las Clases Infames*, 75–77; Harris, *Rebellion on the Amazon*, 182–86, 205–8.

35. Paraense Governor Francisco de Souza Coutinho's official correspondence on the risk of contagion of French ideas, July 8, 1792, in Gomes, De Queiroz, and Coelho, *Relatos de Fronteiras*, 89–91.

36. Anderson, "Following Curupira," 58–60; Cleary, "Lost Altogether," 113–15; Harris, *Rebellion on the Amazon*, chaps. 4 and 6; Hemming, *Amazon Frontier*, 226, 227; Ricci, "De la Independencia a la Revolución Cabana," 55, 81.

37. Harris, *Rebellion on the Amazon*, 205–14.

38. Hemming, *Amazon Frontier*, 229.

39. Smyth and Lowe, *Narrative of a Journey from Lima to Pará*, 300. Andrea is quoted and translated in Cleary, "Lost Altogether," 112.

40. Quoted in Hemming, *Amazon Frontier*, 231.

41. Salles, *O Negro no Pará*, 264.

42. Edwards, *A Voyage up the River Amazon*, 10. See also Lopes, *Emilio Carrey: O Amazonas*, 299–313.

43. Herndon, *Exploration of the Valley of the Amazon*, 346; See also Warren, *Para*, 27.

44. Cleary, "Lost Altogether," 130, in contrast to Harris, *Rebellion on the Amazon*, 280–81.

45. SML—YU, U.S. Consular Reports, Dispatch from Charles Smith, American Consul in Pará, to Sec. of State John Forsyth, September 19, 1835; Charles Smith to Sec. of State John Forsyth, November 7, 1835.

46. Bates, *The Naturalist on the River Amazons*, 36.

47. Herndon, *Exploration of the Valley of the Amazon*, 345.

48. Ricci, "De la Independencia a la Revolución Cabana," 82–85; Salles, *O Negro no Pará*, 265–69; Harris, *Rebellion on the Amazon*, 197.

49. Freitas, *La Revolución de las Clases Infames*, 76; Harris, *Rebellion on the Amazon*, 185.

50. Harris, *Rebellion on the Amazon*, 223.

51. CMA, Lourenço Justiniano de Paiva PMI, 1841, 3–6. Quotations from pp. 2, 23, 25.

52. CMA, Agostinho José Lopes Godinho PMI, 1862; APEP-Orf, Maria Thereza de Jesus de Souza Campos PMI, 1856; CMA, Julio Cezar d'Araujo Danin PMI, 1886. For early 1850s visits of travelers to Danin's and Godinho's properties where both slaveowners talk about the effects of the Cabanagem revolt, see Bates, *The Naturalist on the River Amazons*, 28–29; Edwards, *A Voyage up the River Amazon*, 67–73.

53. Freitas, *La Revolución de las Clases Infames*, 78–79; Lopes, *Emilio Carrey: O Amazonas*, 299–308. Carrey's account is fictional, but is based on primary evidence; Harris, *Rebellion on the Amazon*, 190, 197.

54. For examples from São Paulo and Minas Gerais in 1829, see Klein and Luna, *Escravismo no Brasil*, 180–81. For Chesapeake Bay and the South Carolina Lowcountry in the eighteenth century, see Morgan, *Slave Counterpoint*, 88–90.

55. Barroso, "Coletando o cacau 'bravo,'" 17.

56. Barroso, "Coletando o cacau 'bravo,'" 14. Barroso's sample is from Baixo Tocantins municipalities between 1810 and 1850 and includes some data of the post-Cabanagem period.

57. Kelly-Normand, "Africanos na Amazônia," 17, 19.

58. Barroso, "Coletando o cacau 'bravo,'" 14.

59. Bezerra Neto, *Escravidão negra no Grão-Pará*, 221–22.

60. Pará, *Relatorio . . . 1858*, 34–351; Pará, *Relatorio 1867*, 17. On the number of mills, Cordeiro, *O Estado do Pará*, 20. See Appendx II, Selected Pará Exports.

61. Pará Land Bureau (ITERPA), *Livro de Títulos de Propriedade* no. 1 from Vigia, pp. 10–14, Santo Antônio da Campina Sale Certificate, 1891; CMA, Joaquim Pedro Campos PMI, 1874; APEP-Orf, Anna Joaquina Rosa dos Santos Smith PMI, 1857; Marques, "Modelo da Agroindústria Canavieira Colonial," 111. Many of the thirty-five water mills operating in Pará in 1862 were tide mills: Pará, *Relatorio . . . 1862*, 57–66. See also APEP-Orf, Maria do Carmo do Castilho PMI, 1853; CMA, Antônio Francisco Corrêa Caripuna PMI, 1877. These large sugar plantations received a number of foreign visitors in the post-Cabanagem period: Edwards, *A Voyage up the River Amazon*, 40, 68; Hartt, "A Geologia do Pará," 10–11; Rodrigues, *Rio Tapajós*, 35–43; Spruce, *Notes of a Botanist*, 197–99; Wallace, *Travels on the Amazon and Rio Negro*, 19; Warren, *Para*, 213.

62. Maury, *The Amazon*, 5, 44; Great Britain Foreign Office, *Reports from Her Majesty's Consuls*, 13; Kidder, *Sketches of Residence and Travels*, 2:281; Orton, *The Andes*, 259. For the agricultural ideal, see Costa, *Ecologismo e Questão Agrária*, 7.

63. Kraay, "Slavery, Citizenship and Military Service"; Salles, *O Negro no Pará*, 277–79. On flight, see Toral, "A Participação dos negros escravos," 292.

64. Weinstein, *The Amazon Rubber Boom*, 54; Pará, *Relatorio . . . 1875*, A-XIII.

65. Pará, *Relatorio . . . 1875*, A-XIII. By this time there were beginning to be shortages of other foodstuffs for the same reason, Weinstein, *The Amazon Rubber Boom*, 112.

66. Pará, *Relatorio . . . 1862*, 54.

67. A sample of forty-three postmortem inventories of Paraense sugar mill owners between 1850 and 1880 shows that sugar mills tended to cluster around two ideal models: (1) rum producers with small crews of slaves in their engenhocas and (2) big planters. The fifteen inventories of mills called explicitly *engenhocas* or rum-producing mills had an average of 11.5 slaves per crew—a number close to the 9 workers needed to operate a rum-producing mill in Pará in the 1970s. The second largest group of inventories was that of mill owners with more than forty slaves—the powerful sugar planters discussed earlier. Sample of forty-three sugar planters' inventories from 1850–1880 containing one or more slaves from CMA, APEP, 5A-CR, AFS, and C2A.

68. Data are from Pará, *Relatório . . . 1881*, 130; Lobato, *Caminho de Canoa Pequena*, 99–101, and quotation from Pará, *Relatorio . . . 1875*, 61. See CMA, João de Figueiredo muniz PMI, 1871; CMA, Joaquim Pedro Campos PMI, 1874; APEP-IM, Maria Theresa de Jesus Maia e Miranda PMI, 1876; CMA, Antônio Francisco Corrêa Caripuna PMI, 1877; CMA, Justo José Corrêa de Miranda PMI, 1878.

Chapter Two

1. Lima, "História dos Negros."

2. For more on this narrative, see O'Dwyer, "DaMatta nas paradas"; Ruiz-Peinado, "El Empadronamiento de los Dioses." There is a version narrated by Kaxúyana Indians in Frikel, *Os Kaxúyana*, 12–18. On the 1992 meeting, see Acevedo Marin and Castro, *Negros do Trombetas*, 27–37, 205–49; De la Torre, "Are They Really Quilombos?," 107–8.

3. Lima, "História dos Negros . . . ," 12. On his lineage, see Azevedo, *Puxirum*, 69; Lima, "História dos Negros," 12; Conselho Nacional de Proteção aos Índios and Agricultura, *Diário das Três Viagens*, 13; Funes, "Mocambos do Trombetas," 240. Lima himself and other locals shared their oral traditions with visiting scholars more than once during those years: Ruiz-Peinado, *Cimarronaje en Brasil*, 138–53.

4. Vansina, *Oral Tradition*, 144–46; Portelli, "History-Telling and Time," 60–62; Yow, *Recording Oral History*, 286. I located two other versions of the same narrative, plus other references to the Big Snake, in Azevedo, *Puxirum*, 91–92; O'Dwyer, "DaMatta nas paradas"; Ruiz-Peinado, *Cimarronaje en Brasil*, 138–44; Sauma, "Ser Coletivo, Escolher Individual."

5. On this method, see Frisch, *A Shared Authority*, 22; Ritchie, *Doing Oral History: A Practical Guide*, 85, 101, 119; Vansina, *Oral Tradition*, 160. While not as theoretically sophisticated, my goal is not far from that of Price, *First-Time*, 39–40.

6. Vansina, *Oral Tradition*, 196, 199.

7. Steward, *Handbook of South American Indians*, 3:245–82, 806, 810, 812; Girardi, "Gente do Kaxuru."

8. Reis, *História de Óbidos*, 13–28; Raffles, *In Amazonia*, 75–76.

9. Amaral, *Memorias*, 1:333.

10. Harris, *Rebellion on the Amazon*, 135–40; Roller, "Colonial Routes," chaps. 4 and 5.

11. Maw, *Journal of a Passage*, 341; Harris, *Rebellion on the Amazon*, 160–65. See also Bates, *The Naturalist on the River Amazons*, 122–24; Souza, *O Cacaulista*, 2, 19; Souza, *O Coronel Sangrado*, 113.

12. For example, 24.2 percent in Santarém and 26.7 percent in Alenquer. Bezerra Neto, *Escravidão negra no Grão-Pará*, 225; Salles, *O Negro no Pará*, 286, 299.

13. Brown and Lidstone, *Fifteen Thousand Miles*, 95–98, 115–28, 195–204, 264; Penna, *A região Occidental*, 21, 32; Rodrigues, *Rio Tapajós*, 35–43; Smith, *Brazil, the Amazons and the Coast*, 152–68.

14. Ruiz-Peinado, *Cimarronaje en Brasil*. The phrase "safe haven" is from Gomes, "A Safe Haven." See also Harris, *Rebellion on the Amazon*, 166, 171–72; Roller, "Colonial Routes," 169.

15. Captain of militia Lourenço Justiniano Siqueira to Lower Amazon military commander João Bernardes Borralho, December 24, 1801, in Harris, *Rebellion on the Amazon*, 166.

16. Lima, "História dos Negros," 3, 8; Funes, "Mocambos do Trombetas," 237; Harris, *Rebellion on the Amazon*, 172; Rodrigues, *Exploração e Estudo*, 16; Ruiz-Peinado, "Tiempos Afroindígenas en la Amazonia," 596–97.

17. Rodrigues, *O Rio Trombetas*, 24–26; Harris, *Rebellion on the Amazon*, 168–71; Salles, *O Negro no Pará*, 232–33. On the 1827 expedition, see APEP, Fundo Segurança Pública, Correspondências dos Delegados e Subdelegados, Dionizio Pedro Auzier, Delegado de Policia, to Dr. José Joaquim Pimenta de Magalhães, Police Chief of Pará, January 14, 1854. Document facilitated by José Maia Bezerra Neto. Ibid., 233–34.

18. Ferrer Castro and Acosta Alegre, *Fermina Gómez*, 16–29; Fick, *Making of Haiti*, 58; Rivera-Barnes, "Ethnological Counterpoint," 7–10; Courlander, *A Treasury of Afro-American Folklore*, 94–95, 220–22, 586.

19. Werner, *Myths and Legends of the Bantu*, 59, 49; Brown, *African-Atlantic Cultures*, 112–13, 114–15.

20. Brown, *African-Atlantic Cultures*, 114–15; Weeks, *Among the Primitive Bakongo*, 294.

21. Werner, *Myths and Legends of the Bantu*, 151–53, 231–32.

22. Frikel, *Os Kaxúyana*, 12–16.

23. However, we should not forget that the Big Snake is "the most widespread and well-known supernatural denizen of Amazonian waters." Smith, *The Enchanted Amazon Rain Forest*, 62. See also Rodrigues, *Poranduba amazonense*, 233; Biard, *Deux Années au Brésil*, 350; Frias, *Uma viagem ao Amazonas*, 118–20; Smith, *The Enchanted Amazon Rain Forest*, 62–74.

24. Pará, *Falla . . . 1849*, 22; APEP, *Leis e Decretos da Província do Grão-Pará, 1837–1838*, Law 12, April 25, 1838.

25. On the activities of the Corpo, see AMO, Sessões do Conselho Municipal, 1840–1858, Letters from January 31, 1843, and April 11, 1848; 5A-CR: *Livro de Termos de Conciliação, 1838–1890*, pp. 49–65; Law 330, November 15, 1859, in Gomes, *Nas Terras do Cabo Norte*, 369. On regatões, see Goulart, *O Regatão*, 50–52.

26. Pará, *Exposição . . . 1856*, 108–9; Pará, *Relatorio . . . 1851*, 5; Pará, *Relatorio . . . 1852*, 6. See also Pará, *Relatorio 1855, 15 de Outubro*, 8.

27. Pará, *Exposição . . . 1856*, 3. and AMO, Livro de Sessões do Conselho Municipal, 1840–58, January 1, 1854. See Law 241, December 30, 1853, in Gomes, *Nas Terras do Cabo Norte*, 353.

28. APEP, Secretaria da Presidência da Província do Grão-Pará, Correspondências das Câmaras Municipais, Dionizio Pedro Auzier, Delegado de Policia, to Dr. José Joaquim Pimenta de Magalhães, Police Chief of Pará, January 14, 1854. Document facilitated by José M. Bezerra Neto.

29. A written account of the expedition is discussed in detail and compared to the oral tradition in Ruiz-Peinado, "Maravilla," and in Funes, "Nasci nas Matas," 167–83.

30. APEP, Fundo Segurança Pública, Correspondências dos Delegados e Subdelegados, Dionizio Pedro Auzier, Delegado de Policia, to Dr. José Joaquim Pimenta de Magalhães, Police Chief of Pará, January 14, 1854; APEP, Fundo Segurança Pública, Correspondências dos Delegados e Subdelegados, Alenquer Police Chief to Dr. José Joaquim Pimenta de Magalhães, Pará Police Comissioner, April 25, 1854; Funes, "Nasci nas Matas," 175.

31. Funes, "Nasci nas Matas," 169; APEP, Fundo Segurança Pública, Correspondências dos Delegados e Subdelegados, Alenquer Police Chief to Dr. José Joaquim Pimenta de Magalhães, Pará Police Comissioner, April 25, 1854.

32. Azevedo, *Puxirum*, 78–79; Ruiz-Peinado, "Misioneros en el río Trombetas," 190, 192.

33. Azevedo, *Puxirum*, 78–79; APEP, Fundo Segurança Pública, Correspondências dos Delegados e Subdelegados, Alenquer Police Chief to Dr. José Joaquim Pimenta de Magalhães, Pará Police Comissioner, April 25, 1854; Funes, "Nasci nas Matas," 172.

34. Ruiz-Peinado, "Maravilla," 118. Different species of timbó, such as *Derris ellptica guianensis* (timbó de mata, t. cipó, t. açu, timborana) or *Serjania ichthyctona* (timbó de peixe), are used by Amazonian indigenous and mestiço groups in fishing. Smith, *Brazil, the Amazons and the Coast*, 327.

35. APEP, FSP, CDS, Dionizio Pedro Auzier, Delegado de Policia, to Dr. José Joaquim Pimenta de Magalhães, Police Chief of Pará, January 14, 1854; Letter from Óbidos Police Station, February 9, 1858.

36. Funes, "Nasci nas Matas," 106–7.

37. Santarém Council to the President of the Province, August 9, 1862, quoted in Funes, "Nasci nas Matas," 176; Bastos, *O Vale do Amazonas*, 119–20; Derby, "O Rio Trombetas," 369–70.

38. Brazil, *Recenseamento do Brasil em 1872*, 192; Salles, *O Negro no Pará*, 286.

39. This point is confirmed by Raimundo Vieira dos Santos, another mocambeiro descendant from the Trombetas, in an interview conducted in 1981in Azevedo, *Puxirum*, 68.

40. Funes, "Nasci nas Matas," 297–300, 107; APEP, FSP, CDS, Dionizio Pedro Auzier, Delegado de Policia, to Dr. José Joaquim Pimenta de Magalhães, Police Chief of Pará, January 14, 1854; Rodrigues, *O Rio Trombetas*, 28.

41. See the maroon gender profiles that appear throughout Reis and Gomes, *Liberdade por um Fio*. See also the discussions at Cowling, *Conceiving Freedom*, 51; Miki, "Fleeing into Slavery," 522, 523. For the Caribbean, see Cummings, "Jamaican Female Masculinities," 143–49.

42. Miki, "Fleeing into Slavery," 523.

43. Funes, "Nasci nas Matas," 107–8.

44. Pará, *Falla . . . 1848*, 113; Pará, *Relatorio . . . 1852*, 4; Pará, *Relatorio 1855, 15 de Outubro*, 8; Pará, *Exposição . . . 1856*, 4.

45. Land titling processes from the 1920s and 1930s often detail traditional crops: ITERPA, Autos de Medição "Sucurijú," Raimundo da Costa Lima, 1919; CF, Autos de Medição Judicial de Raimundo da Costa Lima, 1933; ITERPA, Autos de Medição "Tres Barracas" ou "São Braz," Theodora Gonçalves de Lima, 1923; ITERPA, Autos de Medição "Ponta da Gentia," Manoel Costa e Companhia, 1922, and especially ITERPA, São Benedito, Manoel Costa e Companhia, 1923. See also Coudreau and Coudreau, *Voyage Au Trombetas*, 15–19; Maestri and Fabiani, "O mato."

46. Funes, "Nasci nas Matas," 162, 175; Pará, *Falla . . . 1848*, 114; Conselho Nacional de Proteção aos Índios and Agricultura, *Diário das Três Viagens*, 13–14.

47. Pará, *Exposição . . . 1856*, 3; Pará, *Discurso . . . 1845*, 4.

48. Lima, "História dos Negros," 10.

49. Portelli, "History-Telling and Time," 60–62; Portelli, *The Death of Luigi Trastulli*, 70; Vansina, *Oral Tradition*, 45.

50. Batista, "Muito Além dos Seringais," 98–99; Cancela, "Casamento e Relações Familiares," 246; Weinstein, *The Amazon Rubber Boom*, 47–52.

51. AMO, Livro de Sessões do Conselho Municipal, 1840–1858, July 8, 1850, January 11, 1851, January 7, 1873, July 9, 1873; Penna, *Obras Completas*, 2:18–21, 31–32. On governmental support, see Bastos, *O Vale do Amazonas*; Penna, *A região Occidental*.

52. Quotation from Letter from Frei Carmello Mazzarino, January 15, 1868, in Funes, "Nasci nas Matas," 185. See also Ruiz-Peinado, "Misioneros en el río Trombetas."

53. Beatty-Medina, "Between the Cross and the Sword"; Landers, "Cimarrón and Citizen," 124.

54. Ruiz-Peinado, "Misioneros en el río Trombetas," 190–94. Ruiz-Peinado's sources are the interviews conducted with two different Trombetas maroons by Father Guntar Protásio Frikel in 1945. On the fear of reenslavement in the 1860s and 1870s, see Brown and Lidstone, *Fifteen Thousand Miles*, 232; Rodrigues, *O Rio Trombetas*, 35.

55. Kraay, "Slavery, Citizenship and Military Service," 234–35, 243; Toral, "A Participação dos negros escravos," 292.

56. Frei Carmelo de Mazzarino to President Joaquim Raimundo de Lamare, January 15, 1868, in Salles, *O Negro no Pará*, 235.

57. Conselho Nacional de Proteção aos Índios and Agricultura, *Diário das Três Viagens*, 6, 8, 9.

58. Conselho Nacional de Proteção aos Índios and Agricultura, *Diário das Três Viagens*, 11, 13.

59. Coudreau, *Voyage au Cuminá*, 22–36; Coudreau and Coudreau, *Voyage au Trombetas*, 19–22. On Santa Anna, Coudreau, *Voyage au Cuminá*, 18.

60. Conselho Nacional de Proteção aos Índios and Agricultura, *Diário das Três Viagens*, 13, 16, 18, 19; Coudreau, *Voyage à la Mapuerá*, 112; Coudreau and Coudreau, *Voyage au Trombetas*, 23.

61. APEP, FSP, CDS, João Antonio Nunes to José Joaquim Pimenta, January 15, 1854; Police Chief of Óbidos João Baptista Gonsalez Campos to Romualdo de S. Paes d'Andrade, January 9, 1857.

62. AMO, Livro de Atas das Sessões da Camara Municipal, 1858–1872, April 9, 1858; Santarém Camera to President of the Province, August 9, 1862, in Funes, "Nasci nas Matas," 176; Penna, *A região Occidental*, 19, 171.

63. Smith, *Brazil, the Amazons and the Coast*, 327; Rodrigues, *O Rio Trombetas*, 27, 16.

64. Vansina, *Oral Tradition*, 126.

65. On the concept of "insurgent geography," see Miki, "Fleeing into Slavery," 497.

66. Linhares, "Kilombos of Brazil," 831; Treccani, *Terras de Quilombo*, 121; Véran, "Quilombos and Land Rights," 24. Data on demarcations from Andrade, *Quilombola Lands*, 13, 15.

Chapter Three

1. Bates, *The Naturalist on the River Amazons*, 41, 42.

2. Edwards, *A Voyage up the River Amazon*, 40.

3. Edwards, *A Voyage up the River Amazon*, 61; Bates, *The Naturalist on the River Amazons*, 6; Brown and Lidstone, *Fifteen Thousand Miles*, 26; Wallace, *Travels on the Amazon and Rio Negro*, 81; Warren, *Para*, 65.

4. On creolization, see Assunção, *Capoeira*, 32–34; Baron, "Amalgams and Mosaics," 88–95, 112–13; Mintz and Price, *O Nascimento da Cultura Afroamericana*, 33, 76; Parés, "O Processo de Crioulização," 87–97. I considered the environmental side of this process inspired by Carney and Rosomoff, *In the Shadow of Slavery*; Hawthorne, "From 'Black Rice' to 'Brown'"; McNeill, "Envisioning an Ecological Atlantic," 26–27; McNeill, *Mosquito Empires*, 30; Ortiz, *Cuban Counterpoint*.

5. I draw the concept of "landscapes of technology" from Carney, "Landscapes of Technology Transfer."

6. Hawthorne, "From 'Black Rice' to 'Brown'"; McNeill, *Mosquito Empires*, 30; Silver, *A New Face on the Countryside*, 106, 135, 144. Quotation from Assunção, *Capoeira*, 32.

7. Wallace, *Travels on the Amazon and Rio Negro*, 36; Warren, *Para*, 98; Edwards, *A Voyage up the River Amazon*, 47; Spruce, *Notes of a Botanist*, 14. Manioc ovens routinely appear among the objects of slaveholding properties in probate records: CMA, Bernardo José Paes Jr. PMI, 1841; APEP-Orf, De Moraes PMI, 1858; APEP-Orf, Da Silveira PMI, 1864; APEP-Orf, Pastana PMI, 1864; APEP-Orf, De Goes PMI, 1865; CMA, Tusão PMI, 1871.

8. Clement et al., "Origin and Domestication," 76–78; Bates, *The Naturalist on the River Amazons*, 60; Smith, *The Amazon River Forest*, 128–35.

9. Nine of twenty-two cultivars of sweet manioc observed today can be grown in the floodlands. Smith, *The Amazon River Forest*, 129. See Brown and Lidstone, *Fifteen Thousand Miles*, 98; Le Cointe, *L'Amazonie Bresilienne*, 2:117.

10. Meggers, *Amazonia*, 69, 89; Clement et al., "Origin and Domestication," 76.

11. CMA, Lourenço Justiniano de Paiva PMI, 1841; APEP-Orf, Catharina de Sena Maciel Pestana PMI, 1856; APEP-Orf, José Ribeiro de Souza PMI, 1856; and APEP-Orf, Maria Innocência de Nazareth Rodrigues PMI, 1858. Examples of small cultivators appear in Bates, *The Naturalist on the River Amazons*, 43, 100; Anderson, "Following Curupira," 28; Cordeiro, *O Estado do Pará*, 81; Hawthorne, *From Africa to Brazil*, 150.

12. Bates, *The Naturalist on the River Amazons*, 27–28, 38; Edwards, *A Voyage up the River Amazon*, 27–28, 39–42; Kidder, *Sketches of Residence and Travels*, 2:279–80; Warren, *Para*, 197–201, 213. On corn and beans, see Herndon, *Exploration of the Valley of the Amazon*, 299; Wallace, *Travels on the Amazon and Rio Negro*, 63; Edwards, *A Voyage up the River Amazon*, 46; Brown and Lidstone, *Fifteen Thousand Miles*, 97; Lopes, *Emilio Carrey: O Amazonas*, 282–83. See also CMA, Lourenço Justiniano de Paiva PMI, 1841.

13. Lopes, *Emilio Carrey: O Amazonas*, 282; Edwards, *A Voyage up the River Amazon*, 54; Wallace, *Travels on the Amazon and Rio Negro*, 60, 39. See also Smith, *Brazil, the Amazons and the Coast*, 143, 146; Spruce, *Notes of a Botanist*, 11. Fruit trees often appear in probate records as well. See for example APEP-Orf, Carolina Antunes Barral PMI, 1865 and APEP-1V, Felizardo dos Santos Quaresma PMI, 1878.

14. Bates, *The Naturalist on the River Amazons*, 126; Edwards, *A Voyage up the River Amazon*, 51, 135; Biard, *Deux Années au Brésil*, 336, 343; Wallace, *Travels on the Amazon and Rio Negro*, 63, 86; Frias, *Uma viagem ao Amazonas*, 77–79, 126–28.

15. Spix and Martius, *Viagem pelo Brasil*, 74–75; Smith, *Brazil, the Amazons and the Coast*, 308–12, 330–31; Veríssimo, *A Pesca na Amazônia*, 111:56, 76–78, 83.

16. Edwards, *A Voyage up the River Amazon*, 61; Wallace, *Travels on the Amazon and Rio Negro*, 53; APEP-Orf, Maria do Carmo de Castilho PMI, 1853; APEP-Orf, Antônio de Miranda PMI, 1853; APEP-Orf, Anna Raymunda Lobato PMI, 1856.

17. "Venda de Escravo," *Diário do Gram Pará*, August 5, 1870, facilitated by Bezerra Neto; Bezerra Neto, "Escravidão e crescimento econômico," 13–14; CMA, Antônio José Henriques de Lima PMI, 1874; CMA, Manoel Raymundo de Almeida PMI, 1875; CMA, José Joaquim Alves Picanço PMI, 1880; Kingston, *On the Banks of the Amazon*, 467.

18. About two-thirds of the approximately 20,000 enslaved Africans who disembarked in Amazonia came from slave ports located in the coastal strip stretching from Cape Lopez (Gabon), to Benguela (Angola). Bezerra Neto, *Escravidão negra no Grão-Pará*, 216.

19. Chaillu, *Voyages et aventures*, 171; McNeill, "The Ecological Atlantic," 298; Vansina, *Paths in the Rainforests*, 214.

20. Capelo and Ivens, *From Benguella to the Territory of Yacca*, 363–65; Monteiro, *Angola and the River Congo*, 1:287–91.

21. Vansina, *Paths in the Rainforests*, 211–14.

22. Carney, "Landscapes of Technology Transfer"; Carney and Rosomoff, *In the Shadow of Slavery*, 76; Hawthorne, *From Africa to Brazil*, 79.

23. Carney and Rosomoff, *In the Shadow of Slavery*, 55–58; McNeill, "The Ecological Atlantic," 298; Vansina, *Paths in the Rainforests*, 215.

24. Vansina, *Paths in the Rainforests*, 83–86.

25. Bentley, *Life on the Kongo*, 63; Hambly, *The Ovimbundu*, 21:146–48; Capelo and Ivens, *From Benguella to the Territory of Yacca*, 37, 58, 185; Monteiro, *Angola and the River Congo*, 1:97, 179, 183, 209, 211; Vansina, *Paths in the Rainforests*, 214–15.

26. Vansina, *Paths in the Rainforests*, 89. Compare for example to Acevedo Marin, *Julgados da terra*, 174; Maestri and Fabiani, "O mato," 74–79.

27. Capelo and Ivens, *From Benguella to the Territory of Yacca*, 67; Chaillu, *Voyages et aventures*, 251, 438; Vansina, *Paths in the Rainforests*, 91–92.

28. Bentley, *Life on the Kongo*, 63; Chaillu, *Voyages et aventures*, 202; Monteiro, *Angola and the River Congo*, 1:137.

29. The concept of "Indian" used in nineteenth-century Amazonia travel accounts applies to almost any individual living in a palm-thatched hut with a short stature, straight black hair, brown skin, and tribal ornaments, although such individuals often participated in the market economy of the region and did not have a tribal life. If we restrict the definition of Indian to those who lived in a tribal community, then the Indians of the travelogues should better be named "transcultured Indians" or simply "caboclos." For this concept, see Parker, *The Amazon Caboclo*, 1985; Nugent, *Amazonian Caboclo Society*.

30. Warren, *Para*, 222.

31. For the *tipiti*, see Smith, *The Amazon River Forest*, 72, 132–33. For weaving of cotton and other fibers, as well as hammock fabrication, see APEP-Orf, Maria Francisca Ferreira PMI, 1856; APEP-Orf, Ma. Innocencia de Nazareth Rodrigues PMI, 1858; APEP-Orf,

Joanna Antonia dos Santos PMI, 364; APEP-Orf, Marcellina Josefa Ferreira PMI, 1859; APEP-Orf, Maria Thereza de Moraes PMI, 1858; APEP-Orf, Manoel da Cunha Vidinha PMI, 1859, etc.; and Salles, "Memória sobre a rede de dormir." The quotation from Edwards is from *A Voyage up the River Amazon*, 169. See also Fernando L. T. Marques's archaeological studies of water mills in Pará, which show a material culture with both Indian and African cultural elements: Marques, "Modelo da Agroindústria Canavieira Colonial."

32. The Jaguararí engenho, owned by the Jesuits until their expulsion from the Portuguese Empire in 1758, comprised 62 African slaves and 95 Indians in 1761; the Tabatinga cattle farm had 72 Indians and 11 slaves in 1759; and 109 Indians and 16 slaves were residing and toiling in São Caetano in 1759. Marques, "Modelo da Agroindústria Canavieira Colonial," 81, 108; Muniz, "Os Contemplados," 71–78; De Souza Junior, *Tramas Do Cotidiano*, 211, 289–93.

33. Agassiz and Agassiz, *A Journey in Brazil*, 157; Brown and Lidstone, *Fifteen Thousand Miles*, 115, 200, 231; Edwards, *A Voyage up the River Amazon*, 69; Frias, *Uma viagem ao Amazonas*, 237–38; Spruce, *Notes of a Botanist*, 198–99; Wallace, *Travels on the Amazon and Rio Negro*, 63, 82; Warren, *Para*, 213.

34. Champney, *Three Vassar Girls in South America*, 95; Goulart, *O Regatão*, 92–93; Smith, *Brazil, the Amazons and the Coast*, 311–12; Stanfield, *Red Rubber, Bleeding Trees*, 46–62.

35. Wallace, *Travels on the Amazon and Rio Negro*, 80–81; Spruce, *Notes of a Botanist*, 198; Smith, *Brazil, the Amazons and the Coast*, 174; Esch and Roediger, "One Symptom of Originality," 8.

36. Barickman, "A Bit of Land"; Berlin and Morgan, "Introduction"; Carney and Rosomoff, *In the Shadow of Slavery*, 125–27; Forret, *Race Relations at the Margins*, 78–113; Hahn, *A Nation under Our Feet*, 24–25; Mintz, *Caribbean Transformations*.

37. Edwards, *A Voyage up the River Amazon*, 55, 61; Wallace, *Travels on the Amazon and Rio Negro*, 63–64.

38. Berlin and Morgan, "Introduction"; Slenes, *Na Senzala, Uma Flor*, 200–208.

39. APEP-Orf, Marcella Maria de Santa Anna PMI, 1860; CMA, Felipa de Jesús Furtado PMI, 1873. For other slaves engaged in rubber production, which may or may have not implied direct contact with merchants, see notes 16 and 17.

40. APEP-Orf, José Maria de Andrade PMI, 1865, 79, 82–84; AFS, "Mapa de fallida de José Roiz dos Santos Almeida. Diario A," 1870, 3–13; CBA, Caetano José da Costa PMI, 1879. See also APEP-Orf, Cpt. Manoel Hyginio Cardoso Pinto PMI, 1850; APEP-Orf, 1859, D. Anna Theresa de Jesus dos Reys Pessegueiro.

41. AMO, Livro de Sessões do Conselho Municipal, 1840–58, July 8, 1850; Magalhães, *Relatório dos negocios*, 9; Gomes, *Nas Terras do Cabo Norte*, 396, 399.

42. Tocantins, *No tronco da Sapopema*, 40–41; Johann Natterer to Josef Natterer, Borba, December 21, 1829, in Schmutzer, "Der Liebe," 198. See also Edwards, *A Voyage up the River Amazon*, 69–71; Wallace, *Travels on the Amazon and Rio Negro*, 77–78.

43. APEP, Manoel João Corrêa de Miranda PMI, 1870; CMA, Justo José Corrêa de Miranda PMI, 1878. See Acevedo Marin, "Alianças Matrimoniais," 157; Ângelo, "A Trajetória dos Corrêa de Miranda," 27, 31; Batista, "Muito Além dos Seringais," 98.

44. Cancela, "Casamento e Relações Familiares," 259, 264.

45. CMA, Antônio José Henriques de Lima PMI, 1874, pp. 52, 57, 60, 67, 68, 71, 75; Database containing the demographic profiles of 2,138 slaves from rural postmortem inventories with more than one slave between 1850 and 1880, from CMA, APEP, AFS, 5A-V, and CBA.

46. Bolland, "'Proto-Proletarians?,'" 105, 114; Campbell, "As 'a Kind of Freeman'?"; Morgan, "Work and Culture"; Turner, *From Chattel Slaves to Wage Slaves.*

47. CMA, Maria Barbara Gemaque Pereira PMI, 1872; CMA, Antonio José Henriques de Lima PMI, 1874; APEP-Orf, Abric Diniz PMI, 1869, p. 96.

48. Brown and Lidstone, *Fifteen Thousand Miles*, 31, 128. See also Bates, *The Naturalist on the River Amazons*, 87; Biard, *Deux Années au Brésil*, 421; Edwards, *A Voyage up the River Amazon*, 135.

49. Edwards, *A Voyage up the River Amazon*, 40.

50. Florentino and Góes, *A paz das senzalas*, 175–76; Florentino, "The Slave Trade," 301. For a discussion of demographic creolization, see Parés, "O Processo de Crioulização," 88.

51. Teixeira, "Família escrava," 199–200. In the 1870s, Brazil as a whole had a ratio of slave children to mothers of 1,035, compared to 1,056 in the southern United States, considering children ages 1–9 and women ages 15–45. Marcondes, "Fontes censitárias brasileiras," 244; Bergad, "Demographic Change," 924. For a similarly high value of 1,187 in a Paraense plantation, see Barroso, "Múltiplos do Cativeiro," 102.

52. Cowling, *Conceiving Freedom*, 10–11, 57; Toplin, *The Abolition of Slavery in Brazil*, 20–21, 55–59, 92–95.

53. Brazil, *Recenseamento 1872: Pará*, 5:211; Brazil, *Recenseamento do Brasil em 1872*, 3. Even considering the rate of legitimate marriage for individuals older than age 15, it continues to be below the Brazilian average, which itself is not very high. Barroso, "Coletando o cacau 'bravo,'" 18; Klein and Luna, *Slavery in Brazil*, 220.

54. Penna, *Noticia Geral das Comarcas de Gurupá e Macapá*, 26–27; Brown and Lidstone, *Fifteen Thousand Miles*, 82.

55. Subsample of postmortem inventories containing kinship information. It includes forty-two PMIs from 1850–1859 and 635 slaves, or about 2 percent of the total slave population.

56. Klein and Luna, *Slavery in Brazil*, 171; Reis, *A morte é uma festa*, 36.

57. See a discussion at Teixeira, "Família escrava," 184–85.

58. Barroso, "Múltiplos do Cativeiro"; Slenes, "Malungu, ngoma vem!"

59. Marcondes, "Fontes censitárias brasileiras," 244; Barroso, "Coletando o cacau 'bravo,'" 13. Klein and Luna state nonetheless that "the majority of slaves lived in family units." Klein and Luna, *Slavery in Brazil*, 222.

60. Mahony, "Creativity under Constraint," 657. See also Barroso, "Múltiplos do Cativeiro," 125; Mattos, *Das cores do silêncio*, 141, 216; Metcalf, *Family and Frontier in Colonial Brazil*, 155; Morgan, *Slave Counterpoint*, 532, 538–39; Schwartz, *Sugar Plantations*, 379; Slenes, *Na Senzala, Uma Flor*, 48; Stevenson, "The Question of the Slave Female Community," 78.

61. APEP-2V, Autos de Liberdade, A escrava Francisca por seu curador Dr. Augusto Carlos, 1874; see, for example, 5A CR, Autos de Liberdade Lourenço Justiniano da Fonseca, 1873; 5A CR, Autos de Liberdade Jão Maximiano dos Santos, 1874; 5A CR, Autos de Liberdade Manoel d'Assumpção, 1875; 5A-CR, Autos de Liberdade Domingas Maria da Conceição e Filho, 1877.

62. APEP, Juizo Municipal de Breves, Autos Cíveis de Manumissão 1875; APEP-2V, Autos de Liberdade, 1879; in addition, see, for example, APEP 2V, Autos de Liberdade, Raimundo e outros por seu curador, Antonio dos Passos de Miranda, 1879; 5A-CR, Autos de Liberdade, Manoel Zacharias, 1878.

63. Mattos, *Das cores do silêncio*, 216. See also Bolland, "Politics of Freedom," 189–90; Hahn, *A Nation under Our Feet*, 166–70; Mahony, "Creativity under Constraint," 656–57; Penningroth, "The Claims of Slaves and Ex-Slaves," 1063.

64. Hahn, *A Nation under Our Feet*, 21–25; Klein and Luna, *Slavery in Brazil*, 220; Mattos, *Das cores do silêncio*, 151.

65. Harris, *Rebellion on the Amazon*, 78–79, 98; Hecht, "Factories, Forests, Fields and Family," 336–37; Motta-Maués, *"Trabalhadeiras" e "Camarados,"* 22, 45–46. Additional evidence of this emerging labor division is discussed in the next section.

66. Batista, "Demografia, família e resistência," 216.

67. Slenes, *Na Senzala, Uma Flor*, 107.

68. Marques, "Modelo da Agroindústria Canavieira Colonial," 83; CMA, José Ferreira Bello PMI, 1872; ITERPA, Register of the Sale Certificate in the City Council of Vigia on October 28, 1891, Livro de Títulos de Propriedade no. 1, Vigia, pp. 10–14.

69. APEP-IM, Maria Thereza Maia e Miranda PMI, 1876; APEP-Orf, Manoel Raimundo dos Santos Quaresma PMI, 1860.

70. CMA, Antônio Francisco Correa Caripuna PMI, 1877. See also CMA, José Joaquim Pereira da Fonseca PMI, 1872; CMA, Antônio de Souza Monteiro PMI, 1878.

71. On the effects of the 1871 Free Womb Law, see Cowling, *Conceiving Freedom*, 60–62, 76–77.

72. Graham, *Caetana Says No*, 72.

73. CMA, Ma Bárbara Gemaque Pereira PMI, 1872, 99; APEP-Orf, Pedro Baptista de Souza Aranha PMI, 1859; CMA, Antônio José Henriques de Lima PMI, 1874, pp. 52, 57, 60, 67, 68, 71, 75. Database containing the demographic profiles of 2,138 slaves from rural postmortem inventories with more than one slave between 1850 and 1880, from the archives CMA, APEP, AFS, 5A-V, and CBA.

74. Wallace, *Travels on the Amazon and Rio Negro*, 40; Bates, *The Naturalist on the River Amazons*, 87.

75. De la Torre, "O carimbó," 142–44; Amaral and Cordeiro, "Entre homens e mulheres."

76. 5A-CR, Summario de Culpa, Coronel Joaquim Manoel de Carvalho versus Juvenal de Moraes Rego, 1882; Lopes, *Emilio Carrey: O Amazonas*, 198, 201; Edwards, *A Voyage up the River Amazon*, 5, 8–9, 13; Champney, *Three Vassar Girls in South America*, 39–40; Herndon, *Exploration of the Valley of the Amazon*, 391–92; Kidder, *Sketches of Residence and Travels*, 2:294–95; Smith, *Brazil, the Amazons and the Coast*, 42–43; Warren, *Para*, 41, 65. See also Bezerra Neto, "Histórias Urbanas de Liberdade."

77. Bolland, "'Proto-Proletarians?,'" 121.

78. Bezerra Neto, *Escravidão negra no Grão-Pará*, 80–91. The quotation is from p. 91.

79. Klein and Luna, *Slavery in Brazil*, 173–76. The high levels of taxation and high prices for slaves explain this, and the general expansion of the Paraense economy in the rubber era led Paraense elites to retain as many laboring hands as possible. Bezerra Neto, "Escravidão e crescimento econômico," 13–14; Oliveira Filho, "O Caboclo e o Brabo," 128–34; Lacerda, *Migrantes cearenses no Pará*.

80. McNeill, "The Ecological Atlantic," 301; Dean, *With Broadax and Firebrand*, 57.

81. Johnson, "On Agency," 118.

82. By community I mean a local unit where people with a common culture and identity reside, as used by Brazilian historian Sheila de Castro Faria and others: Faria, "Identidade e

comunidade escrava," 145; Pereira de Carvalho, "Autonomia e hierarquia," 23. I emphasize the constructed, experiential nature of the local, as do Lewis, "Connecting Memory," 359–62; Stewart, "Slavery and the Origins," 19–20; Walsh and High, "Rethinking the Concept of Community," 262, 266.

83. Klein and Luna, *Slavery in Brazil*, 220; Mattos, *Das cores do silêncio*, 150–51; Metcalf, *Family and Frontier in Colonial Brazil*, 155; Slenes, *Na Senzala, Uma Flor*, 48.

Chapter Four

1. Hilbert, *A cerâmica arqueológica*, 17; Muniz, *Castanhaes de Alemquer*, 14–17; Secção de Estatistica e Terras da Prefeitura Municipal de Alemquer, *Apontamentos sobre o Município de Alemquer*, 6.

2. Ruiz-Peinado, *Cimarronaje en Brasil*, 160; Funes, "Nasci nas Matas," 268. See also Acevedo Marin and Castro, *Negros do Trombetas*, 210. Quotation from interview with Dona Biquinha (born 1934), Pancada community (conducted in Oriximiná), May 29, 2009. Also interview with Manoel das Graças Pereira (born 1951), Nova Esperança, Erepecú (conducted in Boa Vista), May 25, 2009; interview with Francisco Alegre (born 1952), Boa Vista, June 4, 2009; interview with José Santa Rita dos Santos (born 1922) conducted by Idaliana Marinho de Azevedo in 1988, in Azevedo, *Puxirum*, 44; interview with Raimundo Vieira dos Santos (born 1941) conducted by M. Dutra, in Azevedo, *Purxirum*, 75.

3. Group interview with Nicanor (born 1940), Aldenor Pereira de Jesus (born 1953), Teresa Fernandes Regis (born 1938), and Raymundo Dias Barbosa (born 1947), all from Erepecú Lake, June 6, 2009; "father of the people" or "*pai do povo*" is from interview with Francisco Alegre, June 4, 2009, and also appears in Azevedo, *Puxirum*, 44.

4. For some examples of this mobility, see Lima, "História dos Negros," 4; Sousa, *Diário das Três Viagens*, 13; Funes, "Mocambos do Trombetas," 240. For additional examples, see ITERPA, Process Autos de Medição Leonardo, Raimundo da Costa Lima, 1924, and also at Azevedo, *Puxirum*, 76–77, 82. Muniz, *Indice Geral dos Registros de Terras*, 1:207 contains a number of land claims registered by maroon descendants. I took a conservative approach when cross-referencing names of maroon families in documents and interviews. The following mocambeiro families of the Trombetas River area have extensive family trees: the Dos Santos, De Jesus, Cordeiro, Xavier, Da Silva, Do Carmo, Pereira de Jesus, Da Conceição, Do Espirito Santo, Macaxeira, Regis, and Printes families. In the Erepecurú, Melo, Figueiredo, Oliveira, Almeida, Dos Santos, and Pinheiro frequently occur in the record. I thank anthropologist Julia Sauma for her personal communication on this subject.

5. Coudreau and Coudreau, *Voyage au Trombetas*, 16, 15–19; Ducke, "Explorações Scientíficas," 66–67, 159–60; Cruls, *A Amazônia que eu vi*, 31, 34; Hilbert, *A cerâmica arqueológica*, 2, 10, 17–18. See also Coudreau, *Voyage à la Mapuerá*, 151–52.

6. ITERPA, Autos de Medição "Sucurijú," Raimundo da Costa Lima, 1919; CF, Autos de Medição Judicial de Raimundo da Costa Lima, 1933; ITERPA, Autos de Medição "Tres Barracas" ou "São Braz," Theodora Gonçalves de Lima, 1923; ITERPA, Autos de Medição "Ponta da Gentia," Manoel Costa e Companhia, 1922; ITERPA, São Benedito, Manoel Costa e Companhia, 1923. Coudreau and Coudreau, *Voyage au Trombetas*, 11–16, 19; Coudreau, *Voyage à la Mapuerá*, 150–52; Hilbert, *A cerâmica arqueológica*, 2; Smith, *Brazil, the Amazons and the Coast*, 325–27.

7. Interview with Zé do Carmo (born 1934), Jamary community, May 27, 2009; interview with Maria Rosa Xavier Cordeiro (born 1925), Tapagem community, June 5, 2009; interview with Manoel Francisco Cordeiro Xavier (born 1934), Tapagem community, June 5, 2009. See also interview with Antônio Pereira de Jesus (born 1903), Jamary community, in Azevedo, *Puxirum*, 246; Wagley, *Amazon Town*, 188, 203. The 1931 case is from Cartório Ferreira, Óbidos (henceforth CF), *Justiça Pública v. Philomeno Pinto da Silva, José Lopes, and Raymundo Fragata*, 1931.

8. Coudreau, *Voyage au Cuminá*, 20–22; Coudreau, *Voyage au Rio Curuá*, 19–21. In 1948, anthropologist Charles Wagley also noticed the centrality of the ramada in a black community near Gurupá, Wagley, *Amazon Town*, 30. On Marambiré, see Salles, *Os Mocambeiros*, 79–85; Teixeira, *Marambiré*. On King of Congo ceremonies, see Chasteen, *National Rhythms, African Roots*, 173–75; Fromont, "Dancing for the King of Congo," 200–205.

9. AMO, Livro de Sessões do Conselho Municipal, 1840–1858, January 8, 1850, January 15, 1850, July 5, 1850, and January 11, 1851; AMO, Atas da Cámara Municipal, 1881–1890, July 13, 1889. See also the official or *definitivo* land deeds granted between 1877 and 1940, county of Óbidos (incl. Oriximiná), from ITERPA, *Índice de Títulos Definitivos*; Muniz, *Indice Geral dos Registros de Terras*, 1:207–17.

10. Guimarães, "Rompendo o silêncio," 99; Motta, *Nas fronteiras do poder*, 175–77. On the 1850 and the 1890–1891 land laws, see Dean, "Latifundia and Land Policy"; Ferreira, "Guerra sem fim," 90–173; Holston, "Restricting Access to Landed Property," 122–44; Treccani, *Violência e Grilagem*, 84–85, 93–109.

11. Data about Francisca Maria, Rafael Printes's maternal grandmother, are from Funes, "Mocambos do Trombetas," 253; posse 19,345, in Muniz, *Indice Geral dos Registros de Terras*, 1:213; posse 19,282, registered on June 6, 1898, Muniz, 1:209. There is a high probability that Florencio Antônio dos Santos (posse Santa Maria, number 19,344, p. 213), Margarida Maria de Jesus (posse São Pedro, number 19,484, p. 222), and Raymundo Antônio dos Santos (posse São Raymundo, number 19,520, p. 225) were also mocambeiros; posse 19,344, in Muniz, 1:213.; CF, Demarcação Judicial Raimundo da Costa Lima, 1933; on Frikel's interview see Ruiz-Peinado, "Maravilla," 113. See also posses 19,298 and 19,299, in Muniz, *Indice Geral dos Registros de Terras*, 1:210.

12. Penna, *A região Occidental*, 21; AMO, Livro de Sessões do Conselho Municipal, 1840–1858, 77–87; ITERPA, Petição de compra de José Antônio Picanço Diniz e Manoel Marques Diniz, 1904; properties belonging to Augusto Fonseca Vidal, 1925 (provisional deed from 1923), Francisco de Andrade Figueira, 1929 (1926), Luiz Manfredi, 1925 (1922), Manoel Bentes Soares, 1925 (1920), Margarida Rosa da Conceição, 1927 (1910), Perpetua Monteiro Figueira, 1926 (1923), all from ITERPA, *Índice de Títulos Definitivos*; CF—Inventário Post-Mortem de José Antônio Picanço Diniz 1934; see also Cruls, *A Amazônia que eu vi*, 24–26.

13. Thomaz Antonio D'Aquino in 1861, Father Nicolino José de Souza between 1879 and 1881, engineer Gonçalves Tocantins in 1893, Lieutenant Lourenço Valente do Couto in 1894, Otille Coudreau in 1900, and Avelino de Oliveira and José Antônio Picanço Diniz in 1925; Cruls, *A Amazônia que eu vi*, 81.

14. Muniz, *Castanhaes de Alemquer*, 20.

15. Dean, *Brazil and the Struggle for Rubber*, 46–48; Weinstein, *The Amazon Rubber Boom*, 244–46, 258–59.

16. Municipal Council of Alenquer to the State Government, March 12, 1921, in ITERPA, Martinica property, Antônio Vallinoto, 1920.

17. A hectoliter is 100 liters or approximately 50 kilos of nuts. Almeida, "Do extrativismo à domesticação," 299–300.

18. Frazão, *Castanha do Brasil*, 6; Le Cointe, *L'Amazonie Bresilienne*, 2:462–65.

19. Frazão, *Castanha do Brasil*, 3; Le Cointe, *L'Amazonie Bresilienne*, 2:455–56; Muniz, *Castanhaes de Alemquer*, 54.

20. Peres and Baider, "Seed Dispersal," 604; Salomão, "Densidade, Estrutura e Distribuição," 23, 16–17; Shepard Jr. and Ramirez, "'Made in Brazil,'" 56; Zuidema and Boot, "Demography of the Brazil Nut Tree," 4.

21. Shepard Jr. and Ramirez, "'Made in Brazil,'" 4. It is not possible to produce Brazil nuts outside the region, because the tree depends on a specific type of bee for its pollination and therefore cannot yield the fruit without it. Homma and Menezes, "Avaliação de uma Indústria Beneficiadora," 1.

22. Nineteen deeds, 12 percent of the total, were granted after 1930. Collection of 159 definitive official land deeds granted between 1877 and 1940, county of Óbidos (incl. Oriximiná), from ITERPA, *Índice de Títulos Definitivos*.

23. Cartório Ferreira from Óbidos (henceforth CF), Autos de Manutenção de Posse, *Raymundo da Costa Lima v. João Faria Godinho*, 1928, pp. 2, 17; ITERPA, Autos de Medição Massaranduba, 1923, José Antonio Picanço Diniz (Da Costa Lima was the land surveyor in 1910); Posse Perseverança, José Antonio Picanço Diniz, 1908; Property Vianna Grande, José Gabriel Guerreiro, 1913. See also Emmi, *Oligarquia do Tocantins*; Little, *Amazonia*, 36–39.

24. Jesuino's posse was 19,366 from 1898, Muniz, *Indice Geral dos Registros de Terras*, 1:214; ITERPA, "Sucurijú," RCL, 1919; and "Jacaré," RCL, 1921. It is possible that Araujo was a Portuguese homesteader. Interview with Ruy Brasil (born 1945), Tapagem community, June 7, 2009.

25. Interview with Ruy Brasil (born 1945), Tapagem community (conducted in Oriximiná), June 7, 2009; interview with Anarcindo da Silva Cordeiro (born 1951), Tapagem community (conducted in Oriximiná), June 7, 2009; group interview with Nicanor (born 1940), Aldenor Pereira de Jesus (born 1953), Teresa Fernandes Regis (born 1938), and Raymundo Dias Barbosa (born 1947), all from Lake Erepecú, June 6, 2009.

26. Summary of the process based on the fifty-seven land demarcation files from ITERPA.

27. ITERPA, 5/310 Autos de Medição, Raimundo da Costa Lima, 1917; Autos de Medição Cuicé, Raimundo da Costa Lima, 1923.

28. ITERPA, Autos de Medição Paraíso, Raimundo da Costa Lima, 1921; ITERPA, Autos de Medição Massaranduba, José Antônio Picanço Diniz; Três Ilhas, José Gabriel Guerreiro Júnior, 1925; see also properties Palhal and Mocambinho, both from 1928, belonging to Manoel Costa & Cia; Norte do Rio Craval, 1914, Pedro Martins Dourado; Norte do Cuminámiry, 1914, Antônio Guerreiro de Barros; Rio Erepecurú, 1927, Antonio Pinto da Silva, all from ITERPA, *Índice de Títulos Definitivos*.

29. ITERPA, Autos de Medição Uixal, 1925; Statement of the Buyers, January 25, 1923, in ITERPA, São Benedicto, Manoel Costa & Companhia, 1924.

30. ITERPA, Processo 5/310 Autos de Medição, Raimundo da Costa Lima; "Termo do inicio dos trabalhos de medição e discriminação do lote Fernando," in ITERPA, Autos de medição "Leonardo," Raimundo da Costa Lima, 1924.

31. Raimundo da Costa Lima to the Director of the SLB, ITERPA, Extrema, José Gabriel Guerreiro, 1923; emphasis in the original. Abel Chermont to Director of SLB, September 29, 1924, in ITERPA, Autos de Medição Leonardo, 1924.

32. Abelardo Conduru, Elysio Pessoa de Carvalho's lawyer, to the Director of the SLB, February 2, 1924, at ITERPA, Autos de Medição "Tres Barracas" ou "São Braz," Theodora Gonçalves de Lima, 1923; Applications by Alfredo de Sousa Corrêa and Joaquim Caetano Vianna Gentil, 1910, quoted in Muniz, *Castanhaes de Alemquer*, 21; Coudreau, *Voyage à la Mapuerá*, 159.

33. Interview with Maria de Souza (born 1935), Javary community (conducted in Oriximiná), May 23, 2009; interview with Valério and Zuleide dos Santos (born 1945 and 1955, respectively), May 26, 2009; interview with Dona Biquinha (born 1934), Pancada community, May 29, 2009.

34. For example, Maria Rosa Xavier Cordeiro (born 1925) and her brother Seu Duí (born 1934) started to work at ages 13 and 15, respectively. Group interview with Nicanor (born 1940), Aldenor Pereira de Jesus (born 1953), Teresa Fernandes Regis (born 1938), and Raymundo Dias Barbosa (born 1947), Lake Erepecú, June 6, 2009.

35. Interview with Zé Melo, (born 1942), May 27, 2009; interview with Zé do Carmo, (born 1944), May 27, 2009; interview with Zé Melo (born 1942), May 27, 2009.

36. Interview with Zuleide dos Santos, (born 1955), May 26, 2009. Both CF, Relação Geral dos Bens da Firma Augusto e Emeraldo, 1929, and Petição de Falência da Firma Salon Cohen, 1931, contain abundant information about the products provided by the aviadores and the balances of the extractors. Descriptions of the work performed by the castanheiros may be found at Frazão, *Castanha do Brasil*, 4–5; Le Cointe, *L'Amazonie Bresilienne*, 2:453–57 and at Muniz, *Castanhaes de Alemquer*. I relied on the approximately forty interviews conducted with mocambeiro descendants from Alenquer and Oriximiná, most of them former castanheiros, and on the interviews with Luis Bacellar Guerreiro (born 1929) and Olinda Vallinoto (born 1924), siblings of prominent merchants, as well as a former manager and a former accountant, respectively. A parallel system was employed in the rubber trade: Weinstein, *The Amazon Rubber Boom*, 5–20.

37. Muniz, *Castanhaes de Alemquer*, 39; Borges, *Castanha e Oleaginosas da Amazônia*, 6–8; Frazão, *Castanha do Brasil*, maps "Produção e Exportação" and "Recebedores."

38. Interview with Zé Melo (born 1942), May 27, 2009; interview with Seu Duí (born 1934), June 6, 2009; interview with Luis Bacellar Guerreiro (born 1929), June 9, 2009; ITERPA, Mungubal Property, José Gabriel Guerreiro, 1923; Tucunaré Property, José Gabriel Guerreiro, undated; CF, Autos de Manutenção de Posse, *Elysio Pessoa de Carvalho v. Raimundo da Costa Lima*, 1926, pp. 61–80; CF, Petição de Mandato de Apprehensão de Castanhas, Raimundo da Costa Lima, 1926. See also the list of Brazil nut workers at CF, Relação Geral dos Bens da Firma Augusto e Emeraldo, 1929, pp. 16–17.

39. Interview with Zé Melo (born 1942), May 27, 2009; interview with Dona Biquinha (born 1934), May 25, 2009; group interview with Nicanor (born 1940) and others, June 5, 2009; interview with Maria Rosa Xavier Cordeiro (born 1925), June 5, 2009.

40. Interview with Valério and Zuleide dos Santos (born 1945 and 1955, respectively), May 26, 2009; interview with Zé Melo (born 1942), May 27, 2009.

41. Interview with Maria de Souza (born 1935), May 23, 2009; interview with Maria Rosa Xavier Cordeiro (born 1925), June 5, 2009; interview with Seu Duí (born 1934), June 5, 2009. Azevedo, *Puxirum*, 45.

42. Interview with Antônio Souza (born 1940), Abuí community, June 5, 2009; ANPA, Livro de Petições e Portarias, Município de Alenquer, 1936, Portaria 2 and subsequent Certidão, January 14, 1936, and January 23, 1936; ANPA, Livro de Ocorrências Policiais, Município de Alenquer, 1934, Police Chief to Manoel Rodrigues, March 6, 1930. See also Complaint from Marcos Alves, June 2, 1930, and Police Chief of Alenquer to Subprefeito de Cucuhy, June 4, 1930; APEP, FSP, SCP, Caixa 389, Ofícios recebidos April–May 1925, Arnaldo Pereira de Moraes to Police Chief of Pará, April 4, 1925.

43. APEP, FSP, SCP, Cx 434, Oficios recebidos October–November 1926, Manoel Vicente de Oliveira to State Governor, October 6, 1926; ANPA, Livro de Petições e Portarias do Município de Alenquer, 1933, Portaria 159, July 29, 1935; CF, Manutenção de Posse, *Elysio Pessoa de Carvalho v. Raimundo da Costa Lima*, 1926.

44. Harris, *Life on the Amazon*, 22–23; Harris, *Rebellion on the Amazon*, 40; Nugent, "Whither O Campesinato?," 164–65.

45. Futemma, "The Use of and Access to Forest Resources," 235; Harris, *Life on the Amazon*, 63–64; Nugent, *Amazonian Caboclo Society*, 182–83; Witkoski, *Terras, florestas e águas de trabalho*, 250–51, 289–90.

46. Interview with Seu Duí (born 1934), June 5, 2009; interview with Zé Melo (born 1942), May 27, 2009; evidence substantiating that fear comes from CF, Petição de Manoel Avelino de Oliveira, 1929, requesting that his 11-year-old daughter, who was in José Gabriel Guerreiro's power, be returned to him; Petição de Tutela de Elysio Pessoa de Carvalho, 1930, requesting to be the guardian of his employee Sebastião Vianna Lima's minor daughter; APEP, FSP, SCP, Cx 368, Minutas de Ofícios Enviados, May–July 1924, Police Chief of Pará to Police Sub-Chief of Curuá-Alenquer, June 5, 1924. See also Cowling, *Conceiving Freedom*, 199–213; Guimarães, *Múltiplos Viveres*, 110–12, 157; Rios and Mattos, *Memórias do cativeiro*, 50, 174; Smith, *Brazil, the Amazons and the Coast*, 324.

47. Interview with Seu Duí (born 1934), June 5, 2009; also interview with Antônio Souza and Edith Printes (born in 1940 and 1943, respectively), June 5, 2009; interview with Dona Cruzinha (born 1944), April 20, 2009.

48. "Take revenge" is from Lage, *Quadros da Amazônia*, 50. Interview with Maria José Monteiro or Dona Piquixita (born 1915), Pacoval, April 22, 2009; interview with Dona Maria da Cruz de Assis or Cruzinha (born 1944), Pacoval community, April 20, 2009; interview with Antônio Nâcio Vianna or Tio Nácio (born 1921), May 8, 2009, Pacoval community, conducted in Alenquer.

49. ANPA, Livro de Ocorrências Policiais, Município de Alenquer, 1934, Cpt. Manoel Campello de Miranda to Police Commissioner of Pará, undated (c. September 1931), p. 75; João Petronilla Pereira on behalf of Elysio Pessôa de Carvalho to the Juiz de Direito de Óbidos, November 8, 1923, in ITERPA, Autos de Medição "Tres Barracas" or "São Braz," Theodora Gonçalves de Lima, 1923; the increased overtime is from CF, Manutenção de Posse, *Elysio Pessoa de Carvalho v. Raimundo da Costa Lima*, 1926; Manutenção de Posse, *Raimundo da Costa Lima v. Joào Farias Goudinho*, 1928; ANPA, Livro de Ocorrências Policiais, Município de Alenquer, 1934, cases from March 6, 1930, June 4, 1930, December 29, 1930, February 4, 1931, February 14, 1931, March 24, 1931, etc.

50. Interview with Seu Duí (born 1934), June 5, 2009. See also group interview with Nicanor (born 1940), Aldenor Pereira de Jesus (born 1953), Teresa Fernandes Regis

(born 1938), and Raymundo Dias Barbosa (born 1947), all from Lake Erepecú, June 6, 2009.

51. Statement by José Silva Araujo in Malighetti, "Identitarian Politics," 103–4; statements by Dedé Matos and Lázaro, in Assunção, "A memória," 97, 98. See also Acevedo Marin and Castro, *Negros do Trombetas*, 210; Boyer, "Misnaming Social Conflict," 548; Rios and Mattos, *Memórias do cativeiro*, 50, 121, 204.

52. Smith, *Brazil, the Amazons and the Coast*, 327; Penna, *A região Occidental*, 19, 171; AMO, Livro de Atas das Sessões da Camara Municipal, 1858–1872, April 9, 1858.

53. ITERPA, Autos de Medição "Leonardo," Raimundo da Costa Lima, 1924; Azevedo, *Puxirum*, 77.

54. Interview with Manoel Francisco Cordeiro Xavier (born 1934), June 5, 2009.

55. Interview with José Melo (born 1942), Boa Vista community, May 27, 2009; interview with Francisco Edilberto Figueiredo de Oliveira (born 1958), Poço Fundo community (conducted in Oriximiná), May 29, 2009; interview with João Xavier (Arary), Tapagem community, June 4, 2009 (unknown DOB); interview with Maria Rosa Xavier Cordeiro and Deometilo Cordeiro (born 1925 and 1945, respectively), Tapagem community, June 5, 2009; interview with Seu Duí (born 1934), June 5, 2009; interview with Ruy Brasil, (born 1945), June 7, 2009.

56. Interview with Luis Bacellar Guerreiro (born 1929), Oriximiná, January 9; group interview with Nicanor (born 1940), Aldenor Pereira de Jesus (born 1953), Teresa Fernandes Regis (born 1938), and Raymundo Dias Barbosa (born 1947), June 6, 2009; Interview with Zé Melo (born 1942), May 27, 2009; CF, Manutenção de Posse, *Elysio Pessoa de Carvalho v. Raimundo da Costa Lima*, 1926, pp. 61–62. For a comparable figure for rubber estates, see Esteves, "O seringal," 96.

57. CF, Mandato de Aprehensão de Castanhas, *Raimundo da Costa Lima v. Elyzio Pessoa de Carvalho*, 1926, p. 19; CF, Manutenção de Posse, *Elysio Pessoa de Carvalho v. Raimundo da Costa Lima*, 1926.

58. CF, Manutenção de Posse, *Elysio Pessoa de Carvalho v. Raimundo da Costa Lima*, 1926, 2:70; AMO, Lançamento do Imposto de Industria e Profissão, 1907, commercial houses 197 and 263; AMO, Lançamento do Imposto de Industria e Profissão, 1916, p. 6; Abelardo Leão Conduru, Elysio Pessoa de Carvalho's lawyer, to SLB Director, February 22, 1924, in ITERPA, Autos de Medição "Tres Barracas" or "São Braz," Theodora Gonçalves de Lima, 1923; CF, Manutenção de Posse, *Elysio Pessoa de Carvalho v. Raimundo da Costa Lima*, 1926, 2:61–64.

59. Wagley, *Amazon Town*, 151, 156; Harris, *Rebellion on the Amazon*, 99–103; Heredia, "O campesinato e a plantation," 57; Schwartz, *Slaves, Peasants, and Rebels*, 137–62.

60. Interview with Maria José Monteiro or Dona Piquixita (born 1915), Pacoval, April 22, 2009; Palma Muniz, *Castanhaes de Alemquer*, 28–29; APEP, Documentação Notarial, Livro de Registro de Imóveis de Óbidos, January 9, 1930, sale deed of São Julião, Lake Javary, where the godsons José and Raymundo lived on an adjoining property; see PA, Livro de Batismo 1895–1899, for example, for godparent Rosemiro Marques Baptista on pp. 18, 26, 41, 66.

61. Interview with Dona Cruzinha (born 1944), May 1, 2009; interview with Dona Nazita (Old Vianna's granddaughter, born 1946), Pacoval community, April 21, 2009; interview with Cleinilze Souza Silva (Joaquim Tavares de Souza's granddaughter, born 1954), Santarém, April 27, 2009.

62. Alenquer Parish (or PA), Livro de Batismo 1935–37, pp. 41, 44–47; interview with Dona Nazita (born 1946), April 29, 2009; interview with Cleinilze Souza Silva (born 1954), April 29, 2009.

63. Interview with Seu Duí (born 1934), June 5, 2009; interview with Maria Rosa Xavier Cordeiro and Deometilo Cordeiro (born 1925 and 1945, respectively), June 5, 2009; interview with Ruy Brasil (born 1945), July 7, 2009.

64. CF, Autos de Acção Ordinária de Reconhecimento de Paternidade Cumulada com Petição de Alimentos—Dulce Furtado dos Reis, pelo Assist. Judiciario v. José Anto. Picanço Diniz, 1934. Dulce provided one witness, but her request for recognition of paternity and child support was denied because Diniz accused her of being a prostitute.

65. CF, Relação Geral dos Bens da Firma Augusto e Emeraldo, 1929, p. 15. On sporadic visits of different priests to rural communities, see PA, Livro de Baptismos, 1910–1911; Livro de Óbitos, 1918; Livro de Baptismos, 1935–37. See also group interview with Nicanor (born 1940) and others, all from Lake Erepecú, June 6, 2009.

66. Interview with Francisco Alegre or Colé (born 1952), Boa Vista community, June 4, 2009. Colé worked for Zé Machado, Costa Lima's son-in-law, and his father worked for Costa Lima himself. The phrase "friend of the people" was also used by a maroon descendant in 1988 to refer to the Portuguese owner of a trade store at the Pacoval community, Joaquim Tavares de Souza, in Azevedo, *Puxirum*, 44.

67. Bieber, *Power, Patronage, and Political Violence*, 4, 153; Figueiredo and Silva, "Família, latifúndio e poder"; Frank, "Elite Families," 54–55, 71; Woodard, "Coronelismo in Theory and Practice," 109. According to José Murilo de Carvalho and other authors, this traditional system of domination over the rural population should more aptly be called *mandonismo*: Carvalho, "Mandonismo, Coronelismo, Clientelismo."

68. Raffles, *In Amazonia*, 50, 64–65, 191; Wagley, *Amazon Town*, 91–97, 150–59.

69. Helvécio was both appointed by the governor as mayor and won an election; José Gabriel Jr. was also appointed, and Guilherme won the election twice. Interview with Luis Bacellar Guerreiro (born 1929), June 6, 2009; Tavares, *Inventário Cultural*, 9, 14–15. Raimundo da Costa Lima's Postmortem Inventory, 1941, in ITERPA, Autos de Medição Paraiso, Raimundo da Costa Lima, 1921.

70. For the rest of the state, see Acevedo Marin, "Quilombolas na Ilha de Marajó," 219–22; Gomes, *A Hidra e os Pântanos*, 129–31, 292–98; Pinto, *Nas Veredas*, 66–69. For 1940 data, see Instituto Brasileiro de Geografia e Estatística, *Recenseamento Geral do Brasil . . . 1940*, 144; Wagley, *Amazon Town*, 58–60.

Chapter Five

1. Interviews with Manoel da Conceição de Mello (Seu Nunhes, born 1926, Vigia, March 11), March 31, 2009, and August 10, 2010.

2. Greider and Garkovich, "Landscapes," 1.

3. See for example Acevedo Marin and Castro, *No caminho das pedras*; Castro, *Quilombolas do Pará*; Castro, *Escravos e Senhores de Bragança*.

4. CMA, Agostinho José Lopes Godinho Post-Mortem Inventory, 1862; ITERPA, *Livro de Títulos de Propriedade* no. 1 from Vigia, 13; Edwards, *A Voyage up the River Amazon*, 68–73. Quotations from pp. 68, 71.

5. Information about the tidal mill's functioning are from Seu Nunhes (born 1926), March 31, 2009; interview with Seu Alcides (born 1940), March 26, 2009; interview with Seu Santana (born 1940), March 25, 2009; interview with Ilson Pereira de Mello, aka Seu Cebola (born 1946), March 10, 2009; interview with Dona Bena (born 1922), April 1, 2009; the oxen appear at ITERPA, Register of the Sale Certificate no. 1 from Vigia, 13; CMA, Agostinho José Lopes Godinho Post-Mortem Inventory, 1862; Edwards, *A Voyage up the River Amazon*, 68.; see more on water mills in chapter 1.

6. CMA, Agostinho José Lopes Godinho PMI, 1862; Edwards, *A Voyage up the River Amazon*, 70; the free tenants employed at the sítios also appear at 5A-CR, *O Liberal da Vigia*, September 27, 1877, 11. On the naming of rural properties, see Souza, "Nos Limites da Justiça Paraense," 17–21. On the work spaces of sugar mills in Brazil, see Heredia, "O campesinato e a plantation," 44–47.

7. Interviews with Seu Nunhes (born 1926), March 11 and March 31, 2009; interview with Seu Zacarias (born 1943), April 1, 2009; interview with Dona Bena (born 1922), April 1, 2009. Song no. 79, "Black in the Trunk," Francisco Soeiro, "Collection of 85 Carimbó Lyrics" (Vigia: n.d.). Document provided by Paulo Cordeiro.

8. Quotations from interview with Domingas Moraes (unknown DOB—probably late 1800s), "Raiol na Lembrança de sua Escrava," *O Liberal*, November 14, 1976, and from Tocantins, *No tronco da Sapopema*, 34; interview with Sylvia Helena Tocantins (born 1933), March 3, 2009. On the casa-grande in other states, see Azevedo, *Arquitetura do açúcar*, 96, 106; Heredia, "O campesinato e a plantation," 44; Naro, *A Slave's Place, a Master's World*, 52–55.

9. Interview with Sylvia Helena Tocantins (born 1933), March 3, 2009; interview with Dona Guilhermina (born 1916), March 13, 2009.

10. CMA, Agostinho José Lopes Godinho PMI, 1862; Edwards, *A Voyage up the River Amazon*, 72; on Santo Antônio, see Wagley, *Amazon Town*, 197. Other examples of masters using the blessing in an analogous way are in Edwards, *A Voyage up the River Amazon*, 141; Frias, *Uma viagem ao Amazonas*, 57–58, 63–64; Souza, *O Cacaulista*, 50; Wallace, *Travels on the Amazon and Rio Negro*, 81. On Catholicism and social control in plantations, see Laviña, *Cuba*.

11. On Raiol at this time, see Lima, "Os Motins Políticos," 42–48; Reis, "Motins Políticos," 2.

12. ITERPA, *Livro de Títulos de Propriedade* no. 1 from Vigia, pp. 10–14, Register of the Sale Certificate in the City Council of Vigia on October 28, 1891. A copy of the transcription can be found at Acevedo Marin, *Julgados da terra*, 255–60.

13. The historiography of U.S. slavery has also discussed the gradual merger of community and family among African Americans under slavery, Penningroth, *The Claims of Kinfolk*, 7.

14. Here I am using CMA, Agostinho José Lopes Godinho PMI, 1862 and ITERPA, *Livro de Títulos de Propriedade* no. 1 from Vigia, pp. 10–14, Register of the Sale Certificate in the City Council of Vigia on October 28, 1891. For ethnic diversity as an obstacle to family formation, see Florentino and Góes, *A paz das senzalas*, 35; Parés, "O Processo de Criouliza-ção," 102; Rios and Mattos, *Memórias do cativeiro*, 141.

15. Household censuses in Imperial Brazil show an average of five co-residents with more than 40 percent of households headed by females alone. There was a similar

percentage of female-headed households in the late eighteenth century. Bieber, *Power, Patronage, and Political Violence,* 218n18; Metcalf, *Family and Frontier in Colonial Brazil,* 133.

16. On Raiol's trajectory and family, see Acevedo Marin, "Alianças Matrimoniais"; Cancela, "Casamento e Relações Familiares," 246, 255; Instituto Historico e Geographico do Pará, *Catálogo da primera série,* 36–37; Lima, "Os Motins Políticos," 42; Miranda, *A Família Chermont,* 57–58. Raiol appears as Tavares's attorney at CMA, Agostinho José Lopes Godinho PMI, 1862, p. 24.

17. Edwards, *A Voyage up the River Amazon,* 69–74.

18. Camp, "'I Could Not Stay There,'" 2; Baker, "Introduction," 6; Miki, "Fleeing into Slavery," 497, 511. See also Heath and Lee, "Memory, Race, and Place," 1352–68; Lewis, "Connecting Memory," 347–71.

19. Tocantins, *No tronco da Sapopema,* 63. The eighteenth-century sugar mills at Murutucú, Mocajuba, and Jaguarari did have an L-shaped senzala: Marques, "Modelo da Agro-indústria Canavieira Colonial," 81, 101, 107–8.

20. ITERPA, *Livro de Títulos de Propriedade* no. 1 from Vigia, pp. 10–14, "Register of the Sale Certificate . . . 1891"; CMA, Agostinho José Lopes Godinho PMI, 1862; "Certidão" of the Santo Antônio da Campina property, Cartório do 3º Ofício de Notas de Belém from May 23, 2003, in Acevedo Marin, *Julgados da terra,* 247–54.

21. Similar structures can be seen in Edwards, *A Voyage up the River Amazon,* 54; Frias, *Uma viagem ao Amazonas,* 100; Lopes, *Emilio Carrey: O Amazonas,* 282; Wallace, *Travels on the Amazon and Rio Negro,* 60; Warren, *Para,* 213.

22. Bezerra Neto, "Por todos os meios," 200–201; CMA, Cartório Sarmento 14a Vara Cível—Ação de [unreadable] e nullificação de um papel individual sobre a liberdade do escravo de nome Manoel, 1887. On the *curador de escravos,* see Cowling, *Conceiving Freedom,* 62–63; de la Fuente, "Slave Law and Claims-Making in Cuba," 349.

23. CMA—Ação de nullificação . . . ; Maia Bezerra Neto, "Por Todos os Meios," 209; "No engenho da Campina . . . ," *O Liberal da Vigia,* September 27, 1877; *Diário de Notícias,* September 14, 1884; thanks to José Maia Bezerra Neto for this reference.

24. 5A-CR, Arbitramento dos Escravos Theodoro e Josepha, de Raiol e Irmãos, 1876; Bezerra Neto, "Por todos os meios," 248–49; 5A-CR, Arbitramento dos escravos Theodoro, de Raiol e Irmãos, e Narciza, de Ma Victoria Ferreida de Miranda, 1877; 5A-CR, Arbitramento dos escravos Gonçalo, Carolina, Simplicio, Jeronima e Gregorio (todos filhos de Josepha) de Raiol e Irmão, 1883; APEP—Oficios da Junta Classificadora de Esclavos—Emancipação do escravo Gregório, filho de Theodoro y Josepha, 1884. Established by the Rio Branco Law of 1871, the Emancipation Fund provided funds coming from taxation to buy manumission letters: Bezerra Neto, "Por todos os meios," 318–20; Castilho and Cowling, "Funding Freedom, Popularizing Politics," 95–96, 106; Cowling, *Conceiving Freedom,* 56–57; Toplin, *The Abolition of Slavery in Brazil,* 95–96, 106.

25. Interview with Domingas Moraes (unknown DOB—c. 1888), "Raiol na Lembrança de sua Escrava," *O Liberal,* November 14, 1976. On freed people continuing to reside in former slaveholding properties, see Fraga Filho, *Encruzilhadas da liberdade,* 222–29; Gomes and Machado, "Migraciones, desplazamientos y campesinos negros," 25–31; Guimarães, *Múltiplos Viveres,* 142, 235; Machado, "De rebeldes a fura-greves," 269, 272–79; Mata, "Libertos de Treze de Maio," 173–74; Naro, *A Slave's Place, a Master's World,* 159–61; Rios and

Mattos, *Memórias do cativeiro*, 114–15, 204–11; Rodrigues, "Serra dos Pretos," 173; Stein, *Vassouras*, 267.

26. Bezerra Neto, "Os fundadores de 1917"; Ildone, *Noções de História da Vigia*, 37, 58; Instituto Historico e Geographico do Pará, *Catálogo da primera série*, 36–37; Lima, "Os Motins Políticos," 51–52. In these years he published *Capítulos da História Colonial do Pará* (1894), *Visões do Crepúsculo* (1898), and *Juizo Crítico Sobre as Obras Literárias de Filipe Patroni* (1900). On other fazendas used as summer residences by members of the Brazilian elite, see Figueiredo and Silva, "Família, latifúndio e poder," 1055.

27. Pedro was a state congressman; two of his siblings were engineers and one was a lawyer, and all lived in Rio de Janeiro and São Paulo. FCTN, "Registro Funebre: Baroneza do Guajará," *Folha do Norte*, September 21, 1925; CMA, Pedro Pereira de Chermont Raiol PMI, 1929, p. 5. No memory whatsoever of the baron's sons exists in Cacau in the present. The sale to Campos is at 5A-CR, "Escriptura de compra e venta de terras que o Doutor Pedro Raiol faz à Plinio Wilfrido Campos," April 9, 1928.

28. Gregório is slave no. 41 in the 1874 list, Carolina is 32, and Jerônima is 34. Interview with Dona Guilhermina (born 1916), March 13, 2009; interview with Ana Maria dos Rios (unknown DOB) conducted by Paulo Cordeiro, August 17, 2010; Acevedo Marin, *Julgados da terra*, 118, 129; Cordeiro, *O Carimbó da Vigia*, 24. In response to the widespread absenteeism in the largest sugar mills, other slaves and freedmen worked as overseers in Paraense plantations: Bates, *The Naturalist on the River Amazons*, 87; Biard, *Deux Années au Brésil*, 421; Brown and Lidstone, *Fifteen Thousand Miles*, 31–32, 128; Edwards, *A Voyage up the River Amazon*, 46, 131–35; Frias, *Uma viagem ao Amazonas*, 101; Herndon, *Exploration of the Valley of the Amazon*, 336; Kidder, *Sketches of Residence and Travels*, 2:276; Wallace, *Travels on the Amazon and Rio Negro*, 15.

29. Interviews with Seu Nunhes (born 1926), March 11, 2009, and March 31, 2009. There were similar recollections in the interview with Seu Alcides (born 1940), March 26, 2009, and with Seu Santana (born 1940), March 25, 2009. Seu Nunhes's description of the dam was corroborated by Seu Cebola (born 1946, interviewed on March 10, 2009), by Dona Bena (born 1922, interviewed on April 1, 2009), and by Domingas de Moraes, "Raiol na Lembrança de sua Escrava." Other residents of Mané João included Tio Laudegário, slave no. 51 in the 1874 list, and Alphonse and Emiliana, Ângela's children and nos. 57 and 36. Slave Antônio—Dona Guilhermina and Seu Nunhes's maternal grandparent—is no. 47 (Appendix A). Interview with Dona Guilhermina (born 1916), March 13, 2010; interview with Seu Nunhes by Paulo Cordeiro, August 16, 2010.

30. In the 1874 list, Simplicio was slave no. 55 (Appendix A). Interview with Seu Santana (born 1940), March 25, 2009; interview with Dona Nadi and Seu Ramos (born 1953 and 1948, respectively), March 19, 2009; Acevedo Marin, *Julgados da terra*, 115; in the same list, Carlota is no. 10, Andreza 35, João 56, and Laudegário 51. Gonçalo is no. 54, Vicência 30, Vitorina 31, and Nicolao 52. Interview with Dona Guilhermina (born 1916), March 31, 2009; interview with Seu Nunhes (born 1926), August 17, 2010.

31. On roceiros and posseiros, see Fraga Filho, *Encruzilhadas da liberdade*, 223; Motta, *Nas fronteiras do poder*, 47–48, 74, 78, 144; Motta, "Posseiros no Oitocentos," 85, 99. Interview with Seu Alcides (born 1940), March 26, 2009; many others responded similarly: interviews with Dona Bena (born 1922), April 1, 2009; Nadi Ferreira dos Santos (born 1948), March 3, 2009; Manoel Santana Ferreira (born 1940), March 25, 2009.

32. For example, interview with Seu Vê (unknown DOB), March 10, 2009; interview with Dona Guilhermina (born 1916), March 13, 2009. Also, APEP, Livro de Batismo de Colares, 1895–1898, numbers 187, 190, 200, etc.

33. Tocantins, *No tronco da Sapopema*, 43, 63, 105–8; Ianni, *A luta pela terra*, 61–75; see also 5A-CR, Faustino Nicolau Monteiro PMI, 1874; 5A-CR, Manutenção de Posse, *Antonio Manoel de Belem e Silva v. Miguel Arcangelo da Conceição e outros*, 1880; 5A-CR, *Manutenção de Posse, Serafina dos Anjos Pereira v. Carolino Palheta de Siqueira*, 1905; 5A-CR, Protocolo de Audiências de Colares, April 19, 1910, Termo de Conciliação; 5A-CR, Manutenção de Posse, *Fca. Nery Ferreira v. Antonio Ferreira e outros*, 1911, p.53; 5A-CR, José Barriga PMI, p. 16.

34. On work as domestics and cowboys, interview with Dona Guilhermina (born 1916), March 13, 2009; interview with Sylvia Helena Tocantins (born 1933), March 3, 2009. For fishing, interview with Dona Bena (born 1922), April 1, 2010; Cordeiro, *O Carimbó da Vigia*, 42–44; Motta-Maués, *"Trabalhadeiras" e "Camarados,"* 29–40; Veríssimo, *A Pesca na Amazônia*; interview with Dona Osmarina (unknown DOB, c. 1948), March 10, 2009.

35. Interview with Seu Alcides (born 1940), March 26, 2009; interview with Seu Santana (born 1940), March 25, 2009; Fraga Filho, *Encruzilhadas da liberdade*, 212, 221; Rios, "My Mother Was a Slave," 137–38; Rios and Mattos, "Para além das senzalas," 66–67.

36. Seu Cebola provided ample guidance on the cultivation and processing of manioc. Interview with Seu Cebola (born 1946), March 10, 2009. See also Clement et al., "Origin and Domestication," 76–78; Dean, *With Broadax and Firebrand*, 26–27; Motta-Maués, *"Trabalhadeiras" e "Camarados,"* 45–50; Smith, *The Amazon River Forest*, 128–35; Futemma, "The Use of and Access to Forest Resources," 224–25; Wagley, *Amazon Town*, 65–71.

37. Cordeiro, *O Carimbó da Vigia*, 38–39; Monteiro, *Angola and the River Congo*, 1:300. See also Naro, *A Slave's Place, a Master's World*, 5.

38. For a general introduction to carimbó, see IPHAN, *Dossiê: Carimbó*; Salles and Salles, "Carimbó."

39. Salles, *O Negro no Pará*, 187; Cordeiro, *O Carimbó da Vigia*, 21; 5A-CR, "A propósito de S. Caetano," *O Espelho*, January 12, 1879; APEP, Municipal Laws of Vigia, 1883, Article 48, paragraph 2; "Raiol na Lembrança de sua Escrava," *O Liberal*, November 14, 1976.

40. Cordeiro, *O Carimbó da Vigia*, 40, 46–47; Amaral and Cordeiro, "Entre homens e mulheres," 148–49; Palheta, *Vigia Ainda Ontem*, 79–80. On the etymology and the origins of carimbó and zimba see IPHAN, *Dossiê: Carimbó*, 14; Miranda, *Glossario paraense*, 23; Salles and Salles, "Carimbó," 259, 263, 278–82. In 2013, the Brazilian Institute of Historic and Artistic Patrimony or IPHAN declared carimbó part of the Brazilian Cultural Patrimony, and recognized Tauapará as one of the regions where it originated, IPHAN, *Dossiê: Carimbó*, 96.

41. Interview with Dona Nadi (born 1953), March 19, 2009; see interview with Seu Santana (born 1940), March 25, 2009; interview with Seu Zacarias (born 1943), April 1, 2009.

42. "Raiol na Lembrança de sua Escrava," *O Liberal*, November 14, 1976. Quote is from Andrews, *Blackness in the White Nation*, 27. The lyrics are from songs 18, 6, 3, 42, and 28 in Francisco Soeiro, "Collection of 85 Carimbó Lyrics" (Vigia: c. 1960). Document provided by Paulo Cordeiro. See also Salles and Salles, "Carimbó." On the *morena*, see songs number 16, 60, 58, 54, 26, 57 from Soeiro, "Collection of 85 Carimbó Lyrics." See also Chasteen, *National Rhythms, African Roots*, 201–4.

43. Song 79 from Soeiro, "Collection of 85 Carimbó Lyrics"; Heath and Lee, "Memory, Race, and Place," 1356; Tilley, "Introduction," 17–18; Raffles, *In Amazonia*, 23.

44. Song 79 from Soeiro, "Collection of 85 Carimbó Lyrics"; Brown, *The Reaper's Garden*, 231; Tilley, "Introduction," 21. In 2009, I argued that the famous *Dança da Onça* or Jaguar Dance refers to slavery: De la Torre, "O carimbó."

45. For "places of memory," see Portelli, "History-Telling and Time," 60–62; Portelli, *The Death of Luigi Trastulli*, 70; Rubert and Silva, "O acamponesamento como sinônimo," 270. Quotes from Baker, "Introduction," 6; Camp, "'I Could Not Stay There,'" 2; Miki, "Fleeing into Slavery," 511.

46. Interview with Dona Guilhermina (born 1916), March 13, 2009.

47. Interview with Seu Zacarias (born 1943), April 1, 2009; interview with Seu Alcides (born 1940), March 26, 2009. The "map" also appeared in interviews with Seu Nunhes (born 1926) on March 11 and on March 31, 2009, and in an interview conducted by Paulo Cordeiro with Seu Nunhes, August 10, 2010. It probably originated the belief that the lands of Cacau had been donated by the baron to his slaves: "Raiol na Lembrança de sua Escrava," *O Liberal*, November 14, 1976; interview with Francisco Soeiro by Almeida, *Tauaçará*, 60; Palheta, *Vigia Ainda Ontem*, 79. For explorations of how real and imagined documents and rituals produce real ownership, see Besson, "The Appropriation of Lands"; Blomley, "Landscapes of Property"; Milner, "Ownership Rights and the Rites of Ownership."

48. Interview with Dona Guilhermina (born 1916), March 13, 2009.

49. Interview with Seu Cebola and Seu Nunhes conducted by Paulo Cordeiro, August 10, 2010.

50. Interview with Sylvia Helena Tocantins (born 1933), March 3, 2009; Tocantins, *No tronco da Sapopema*, 81–95.

51. Oliveira Filho, "O Caboclo e o Brabo," 121–25; Harris, *Life on the Amazon*, 23–31; Harris, *Rebellion on the Amazon*, 51–61; Parker, "The Amazon Caboclo," 1985, 33; Weinstein, "Persistence of Caboclo Culture in the Amazon," 99, 105; Weinstein, *The Amazon Rubber Boom*, 38, 157.

52. Fried egg and manioc flour form a basic lunch in the region. Interview with Seu Cebola (born 1940), March 19, 2009.

53. Interview with Seu Nunhes (born 1926), March 11, 2009. This could correspond to the *terça* requested from black tenants in other Brazilian states: Mattos, "Políticas de reparação e identidade coletiva no meio rural," 173; Moura, *Os deserdados da terra*, 81.

54. All fragments from interviews with Seu Nunhes (born 1926) from March 11 and March 31, 2009, and August 10, 2010.

55. APEP—Fundo do Executivo, Repartição de Obras Públicas, Terras e Viação—Série 8: Ofícios ROPTV-Intendências. Municipal Council of Colares to the State Governor, October 20, 1894; APEP—F.E., ROPTV—S.8. Municipal Council of Colares to the Director of Public Works, Land, and Colonization, October 20, 1894; Colares's patrimonial land in Muniz, *Indice Geral dos Registros de Terras*, 1:64. See also Muniz, *Patrimonios dos Conselhos Municipaes*.

56. "O título deste Conselho ... sobreexiste e preexiste a todos," APEP—F.E., ROPTV—S.8. Municipal Council of Colares to the Director of Public Works, Land, and Colonization, October 30, 1894; On sesmarias, see Motta, *Nas fronteiras do poder*, 120–32.

57. "Allegam cultivo efetivo e morada habitual n'essas áreas de terras, posse mansa e pacífica, e mil cousas a que se podria rezar," APEP—F.E., ROPTV—S.8. Letter . . . October 30, 1894.

58. For an analysis of the concept of posse in relation to Brazilian agrarian history, see Brannstrom, "Producing Possession," 871; Harris, *Life on the Amazon*, 78; Holston, "Restricting Access to Landed Property," 121–22; Meira, "Aquisição da propriedade pelo usucapião"; Motta, *Nas fronteiras do poder*, 123–27.

59. Quotation from Motta, "Posseiros no Oitocentos," 94; Dean, "Latifundia and Land Policy," 618–25; Ferreira, "Guerra sem fim," 132, 173; Holston, "Restricting Access to Landed Property," 131–44; Motta, *Nas fronteiras do poder*, 175; Naro, "Customary Rightholders and Legal Claimants," 491–501; Paoliello, "As comunidades tradicionais no Vale do Ribeira," 77.

60. Anderson, "Following Curupira," 226–27; Brannstrom, "Producing Possession," 871, 877; Garcia, "Senhores de terra e intrusos," 144–49; Paoliello, "As comunidades tradicionais no Vale do Ribeira," 77.

61. 5A-CR, Ação Cível de Manutenção de Posse, *Irmãos Pinheiro v. Silvestre Sarmento e sua mulher, José Manoel dos Santos e sua mulher*, 1905, p.71. Some examples: 5A-CR, Ação Cível de Interdicto Possessório, *Pinheiro v. Dos Santos*, 1872; 5A-CR, Manutenção de Posse, *António Manoel de Belém Silva v. Manoel Thomas do Nascimento*, 1880; 5A-CR, Autos de Demarcação Judicial, Honório José dos Santos e Dona Ignácia, 1884; 5A-CR, Manutenção de Posse, *De Macedo v. De Souza*, 1905; 5A-CR, Ação Cível de Embargo, Vários Autores e Réus, 1908; 5A-CR, Manutenção de Posse, *Francisca Nery Ferreira v. Antonio Ferreira e outros*, 1911; Cartório Ferreira, Óbidos, Ação Cível de Reivindicação de Posse, *Verônica Florisberta Torres de Souza v. José Gonçalves Guimarães e sua mulher Constantina Pinto Gonçalves*, 1920; APEP—Fundo Segurança Pública—Chefatura de Polícia, Ofícios Diversos—Caixa 339—Letter from Antônio Motta de Souza Neves to Police Chief of Pará, November 3, 1922; APEP—FSP—CP,OD—C 339—Letter from Raymundo Moreira de Souza to Police Chief of Pará, November 4, 1922; APEP—FSP—CP,OD—C 339—Manoel Pereira de Andrade to Police Chief of Pará, November 11, 1922; APEP—FSP—CP,OD—C 373—José Lino to Police Chief of Pará, October 6, 1924, and its reply in APEP—FSP—CP,OD—C 375—Letter from Police Chief of Pará to Security Deputy of Santarem, October 7, 1924; APEP—FSP—CP,OD—C 374—Letter from Antônio Victor de Almeida to Police Chief of Pará, October 6, 1924, and its reply in APEP—FSP—CP,OD—C 375—Letter of Police Chief of Pará to Security Deputy from Acará, October 7, 1924.

62. Garcia, "Senhores de terra e intrusos," 151; Guimarães, "Rompendo o silêncio," 108; Motta, *Nas fronteiras do poder*, 196–211; Naro, "Customary Rightholders and Legal Claimants," 512–16.

63. 5A-CR, Autos de Embargo, *Agostinho José Lopes Godinho v. Favacho e outros*, 1857, p. 26; 5A-CR, *Autos de Embargo Agostinho José Lopes Godinho v. João de Medeiros e outros*, 1857. Lands attached to convents also generated traditional communities all over Brazil: Almeida, *Tierras tradicionalmente ocupadas*, 111–13.

64. Domingos Antônio Raiol, *Motins Politicos ou Historia dos Principaes Acontecimentos Politicos da Provincia do Pará desde o Anno de 1821 até 1835*, Vol. 4, (Rio de Janeiro: Typ. Hamburgueza do Lobão, 1884), 349–51.

65. Still, a number of historians have noticed recently how using the legal system to evict posseiros was "a double-edged sword" and "did not guarantee [the landowners] a victory": Brannstrom, "Producing Possession," 871; Guimarães, "Rompendo o silêncio," 108, 111, 123; Motta, "Posseiros no Oitocentos," 96; Naro, "Customary Rightholders and Legal Claimants," 516. Quotations from Motta, *Nas fronteiras do poder*, 104; Garcia, "Senhores de terra e intrusos," 128–29.

66. Recent analyses of peasant populations in Brazil have noted the centrality and the political nature of this struggle: Garcia, "Senhores de terra e intrusos," 151; Guimarães, "Rompendo o silêncio," 123; Motta, "Classic Works," 137; Motta, *Nas fronteiras do poder*, 211; Paoliello, "As comunidades tradicionais no Vale do Ribeira," 77; Paoliello, "Condição camponesa," 230. Quotations from Motta, "Posseiros no Oitocentos," 99; Naro, *A Slave's Place, a Master's World*, 182. Other historians have emphasized the political nature of claims over land and labor in peasant societies: Hahn, *A Nation under Our Feet*; Mallon, *Peasant and Nation*; Scott, "Reclamando la Mula de Gregoria"; Scott and Zeuske, "Demandas de Propiedad y Ciudadanía."

67. Greider and Garkovich, "Landscapes," 2; Tilley, "Introduction," 17–18.

68. Document SR-01/PA 54100.000111/2005-30 from INCRA, the Brazilian Institute for Agrarian Reform, granting the quilombo desendant communities of Cacau and Ovos 33,123 hectares. In that year forty-four families were registered as living there, curiously the same number of them listed in Seu Nunhes's map from 1943. Accessed January 5, 2016. http://laced .etc.br/site/sistema_quilombo/comunidade.php?idQuilombo=1355.

Chapter Six

1. On Brazil nuts, see chapter 4, note 19.

2. On the origins and history of Pacoval, see Funes, "Nasci nas Matas."

3. APEP, Executive Power Fund, Muniz, *Castanhaes de Alemquer*, 16–17, 60. Emphasis is from the original unless otherwise stated.

4. Muniz, *Castanhaes de Alemquer*, 46, 57.

5. Muniz, *Castanhaes de Alemquer*, 16; APEP, FSP, SCP, Cx 409, Oficios recebidos February 1925–January 1926, Commissioner Manoel Roberto de Azevedo Vasconcellos to the Óbidos Police Chief, December 28, 1925. Alberto, *Terms of Inclusion*, 53–64; Andrews, *Afro-Latin America*, 128, 139–42; Butler, *Freedoms Given, Freedoms Won*, 86, 94, 110, 121; Ferrara, *A Imprensa Negra Paulista*, 51, 105.

6. Nugent, *Amazonian Caboclo Society*; Parker, *The Amazon Caboclo*, 1985. The concept has been heavily criticized because of its pejorative connotations and the fact that it seems to be externally imposed: Harris, "What It Means to Be Caboclo"; Pace, "The Amazon Caboclo."

7. On the economy and the history of Alenquer, see Instituto Brasileiro de Geografia e Estatistica, "Alenquer—PA"; Secção de Estatistica e Terras da Prefeitura Municipal de Alemquer, *Apontamentos sobre o Município de Alemquer*, 1–3, 6–8, 10–12; Simões, *Municipio de Alemquér*, 65–77.

8. AMO, Avulsos, Intendencia Municipal de Alemquér to Óbidos Mayor, March 17, 1917. See also Intendencia Municipal de Óbidos to Governor of Pará, January 5, 1917

9. Alenquer City Council to State Government, March 12, 1921, at ITERPA, Property Martinica, Antônio Vallinoto, 1920.

10. Henrique Santa Rosa to State Secretary General, April 12, 1921, at ITERPA, Property Martinica, Antônio Vallinoto, 1920; FCTN, "Castanhaes de Alemquer," *Folha do Norte*, July 12, 1921; Muniz, *Castanhaes de Alemquer*, 9.

11. Muniz, *Indice Geral dos Registros de Terras*; Muniz, *Legislação de Terras: Dados Estadísticos*; Muniz, *Limites Pará-Goyaz: Notas e Documentos*; Muniz, *Patrimonios dos Conselhos Municipaes*. Among his numerous historical studies the most important were Muniz, "Adhesão do Grão-Pará à Independência"; Muniz, *Estado de Grao-Pará*; Muniz, "Grenfell na história do Pará, 1823–1824"; Muniz, *O Instituto Santo Antonio do Prata*; Muniz, "Os Contemplados."

12. Muniz, *Castanhaes de Alemquer*, 12, 14, 16–17. See also APEP, Fundo do Executivo, SOPTV, Palma Muniz to State Secretary, August 25, 1921.

13. On maroons in this region, see Acevedo Marin and Castro, *Negros do Trombetas*; Arregui, "Amazonian Quilombolas"; Azevedo, *Puxirum*; Funes, "Mocambos do Trombetas"; Funes, "Nasci nas Matas"; Gomes, *A Hidra e os Pântanos*, 43–128; Gomes, "A Safe Haven"; Gomes, "Etnicidade e Fronteiras Cruzadas nas Guianas (sécs. XVIII–XX)"; O'Dwyer, "Os Quilombos"; Ruiz-Peinado, "Amazonía Negra"; Ruiz-Peinado, *Cimarronaje en Brasil*; Ruiz-Peinado, "Tiempos Afroindígenas en la Amazonia"; Larrea and Ruíz-Peinado, "Memoria y Territorio Quilombola en Brasil"; Salles, *O Negro no Pará*, 231–39; Salles, *Os Mocambeiros*.

14. Pará, *Exposição . . . 1856*, 3, and AMO, Livro de Sessões do Conselho Municipal, 1840–1858, January 1, 1854.

15. The exact numbers are unknown. Funes, "Nasci nas Matas," 97, 193–226. See also Teixeira, *Marambiré*, 23–26.

16. Funes, "Nasci nas Matas," 227–28; JRPC, Interview with Aureliano de Sousa Castro (born in 1903) by José Ubaldo de Oliveira Reis, Alenquer, August 7, 1992; Law 751 from February 25, 1901, in FCTN, "Congresso do Estado: Senado," *Folha do Norte* 9522, September 23, 1921; Muniz, *Castanhaes de Alemquer*, 47.

17. There is evidence that a clandestine trade of nuts between maroons and merchants had existed since at least the 1860s: APEP, FSP, Série Chefatura de Polícia, João Antonio Nunes to José Joaquim Pimenta, January 15, 1854; Police Chief of Óbidos João Baptista Gonsalez Campos to Romualdo de S. Paes d'Andrade, January 9, 1857; for the 1870s and 1880s, see CB, Luiz de Oliveira Martins PMI, 1881; C2A, Candido José Simões PMI, 1873, p. 5; C2A, Manoel Pinto Monteiro PMI, 1887; see also Azevedo, *Puxirum*, 30; Funes, "Nasci nas Matas," 231–32. Quotation from Santarém Camera to President of the Province, August 9, 1862, in Funes, p. 176.

18. Cartório do 2° Ofício, Luiz de Oliveira Martins PMI, 1881; Cartório do 2° Ofício ("Toninho"), Alenquer, henceforth C2A, Candido José Simões PMI, 1873, p. 5; C2A, Manoel Pinto Monteiro PMI, 1887; Azevedo, *Puxirum*, 30; Funes, "Nasci nas Matas," 231–32.

19. See narrative gathered in 1945 by Protásio Frikel, in Funes, "Nasci nas Matas," 170–72, 176; Ruiz-Peinado, "Maravilla," 115.

20. Muniz, *Castanhaes de Alemquer*, 39–44.

21. On compadrio and its iterations in Amazonia and beyond see Bieber, *Power, Patronage, and Political Violence*, 74; Diacon, "Peasants, Prophets," 509–13; Esteves, "O seringal," 99; Heredia, "O campesinato e a plantation," 56–57; Moreira, "Memória Social," 122; Raffles, *In Amazonia*, 50, 57, 63; Rubert and Silva, "O acamponesamento como sinônimo," 271; Wagley, *Amazon Town*, 151–56.

22. Funes, "Nasci nas Matas," 231; Muniz, *Castanhaes de Alemquer*, 19. Áurea Nina, "Loanda, Sua História, Sua Luta . . . ," Unpubl. thesis. Santarém: UFPA, 1993, 14; JRPC, interview with Aureliano de Sousa Castro (born 1923), by José Ubaldo Oliveira Reis, Alenquer, August 7, 1992.

23. ANPA, Interview with José Rafael Valente, (born 1912), Alenquer, 1992; interview with Maria da Cruz de Assis (born 1944), Pacoval, April 20, 2009; Coudreau, *Voyage au Rio Curuá*, 22–23; Funes, "Nasci nas Matas," 301–5.

24. He was Pará's state attorney, legal advisor to the State Bureau of Public Lands, and attorney of the State Treasury. FCTN, "Quem Foi Fulgêncio Simões," *O Surubiú*, Alenquer, August 26, 1999; ICPA, "Dr. Fulgencio Simoes," *Alenquerense: Folha Mensal* 13, August 30, 1942, 1; C2A, Antônio Firmino Simões PMI, 1893, pp. 7–8, 13–14.

25. Law 751 from February 25, 1901, at FCTN, "Congresso do Estado: Senado," *Folha do Norte* 9522, September 23, 1921. Apparently the proposal did not succeed. The *Diário Oficial do Estado do Pará* did not list it, and nobody in Pacoval had any knowledge of this law. His linkages to Pacoval seem to come from his father's relations with Luiz de Oliveira Martins and from his operations in the Curuá River area: C2A, Antônio Firmino Simões PMI (henceforth PMI), 1893, pp. 7–8, 13–14; IHGP-APM, "Nona Sessão Ordinaria," Câmara de Municipal da Vila de Alenquer, Livro de Atas 1855–1862, p. 12.

26. JRPC, interview with Everaldo Antonio de Jesus (born 1923) conducted by José Ubaldo Oliveira Reis, Alenquer, July 7, 1992; interview with Raimundo Francisco Cardoso (born 1907) conducted by José Ubaldo Oliveira Reis, Alenquer, July 28, 1992.

27. Coudreau, *Voyage au Rio Curuá*, 19. This narrative overlapped with another one arguing that the lands of Pacoval were granted by Brazilian emperor Dom Pedro II as a *sesmaria* or royal land grant. Sesmarias were abolished in the 1820s, but here the Pacovalenses seem to be mixing in elements from their 1877 forced journey to Belém, when one of them was liberated by the emperor. Azevedo, *Puxirum*, 30.

28. Muniz, *Castanhaes de Alemquer*, 46.

29. Aliprandi and Martini, *Gli Italiani del Nord del Brasile*, unnumbered page. Olinda Vallinoto, "Recordando Minhas Raízes," mimeo; interview with Olinda Vallinoto (born c. 1916), Alenquer, April 20, 2009.

30. AMO, Lançamento do imposto de industria e profissão, 1916; Lançamento do imposto de industria e profissão, 1918. See also Emmi, *Italianos Na Amazônia*. For the Portuguese, see Joaquim Tavares de Souza's Personal Diary (1893–1897), pp. 5 and 6 (undated, c. 1894), provided by Cleinilze Sousa e Silva, Alenquer, April 27, 2009. Thanks to her for this invaluable document. Carícia de Sousa e Silva, "Um Pouco de Nossa História: Avô Materno," mimeo provided by Carícia, April 2009; APEP, Fundo Secretaria de Segurança Pública, Caixa 528, Chefatura de Polícia, Petições, Janeiro-Março de 1933, João Miléo Primo to State Police Chief, March 3, 1933; see adjacent years.

31. APEP, FSP, SCP, Cx 409, Oficios recebidos February 1925–January 1926, Commissioner Manoel Roberto de Azevedo Vasconcellos to the Óbidos Police Chief, December 28, 1925.

32. APEP, FSP, SCP, Cx 409, Oficios recebidos February 1925–January 1926, Commissioner Manoel Roberto de Azevedo Vasconcellos to the Óbidos Police Chief, December 28, 1925. For incidents where the mocambeiros used a similar vocabulary, see CF, Ação Sumária de Esbulho, *Manoel Costa v. Irineu da Silva Pinto e outros*; ITERPA, Autos de

medição "Leonardo," Raimundo da Costa Lima, 1924; Coudreau, *Voyage au Rio Curuá*, 19, 22–23.

33. APEP, FSP, Chefatura de Polícia, Ofícios Diversos, Fevereiro–Março 1923, Cx 349, Juruty Police Chief to State Police Chief, March 14, 1923. See also AMO, Registro de Ofícios, 1920–1930, Óbidos Council to Fiscal Post in Igarapé-Assu, March 22, 1923.

34. APEP, FSP, Oficios Recibidos Dezembro 1925–Janeiro 1926, Cx 407, Deodoro Mendonça to State Chief of Police, November 30, 1925. The letter to the Italian consul was signed by Biagio Mileo, Biagio Calderaro, Vincenso Sarubi, Biagio Sarubi, Nicola Mileo, Pasquale Sarubi, Nicola Ferrari, Giuseppe Mileo, Pellegrino Costabile, Pietro Oliva, Biagio Belle, Giuseppe Calderaro, Antonio Calderaro, Caino Antonio, and Guiseppe Sarubbi.

35. APEP, FSP, Oficios Recibidos Dezembro 1925–Janeiro 1926, Cx 407, Antonio Machado Imberiba to State Chief of Police, December 16, 1925.

36. Muniz, *Castanhaes de Alemquer*, 57.

37. Chambouleyron, *Povoamento, Ocupação e Agricultura*, 101; Costa, *Ecologismo e Questão Agrária*, 3–9; Costa, "Lugar e significado," 187; Harris, *Rebellion on the Amazon*, 52, 111, 122, 136–37; Weinstein, *The Amazon Rubber Boom*, 37–40.

38. Pará, *Mensagem . . . 1918*, 15–22.

39. Pará, *Mensagem . . . 1920*, 80; Pará, *Mensagem . . . 1919*, 85. On Sodré, see Borges, *Vultos Notáveis do Pará*, 204–12; Sodré, *Lauro Sodré*.

40. Pará, *Mensagem . . . 1925*, 14–15.

41. "O jubileu profissional do engenheiro Henrique Santa Rosa," *Folha do Norte* 12,823, February 10, 1921; Weinstein, *The Amazon Rubber Boom*, 106–7.

42. This is because different bees need to pollinate its flowers in order to produce the seeds (which become the "nuts"). Mori and Prance, "Taxonomy, Ecology," 138–41.

43. Quotations from Henrique Santa Rosa, "Regimen de Terras do Estado," in Pará, *Mensagem . . . 1925*, 107, 109. Exactly the same position is defended in S. Torres Videla, "A Desobstrução dos Cursos d'Agua," *Folha do Norte*, Jauary 1, 1925. See also Pará, *Mensagem . . . 1927*, 108.

44. Santa Rosa, "Regimen de Terras," Pará, *Mensagem . . . 1925*, 107. Carvalho, *The Formation of Souls*, 15–16, 24; Resende, "O processo político," 103–6. Sodré himself had studied in the late 1870s at the military academy of Praia Vermelha, a stronghold of positivist thought: Sodré, *Lauro Sodré*, 15–17.

45. Muniz, *Castanhaes de Alemquer*, 57.

46. Carvalho, *The Formation of Souls*; Resende, "O processo político"; Woodard, *A Place in Politics*.

47. Love, *The Revolt of the Whip*, 14–20; Muniz, *Indice Geral dos Registros de Terras*, 1:xx; quotation from Woodard, *A Place in Politics*, 52.

48. While it has not been totally proven that he was the author of that quote, it nonetheless expresses very well the ideas about poverty that republican politicians often held. Carvalho, *The Formation of Souls*, 22–23, 151; Fausto, "Dos governos militares a Prudente-Campos Sales," 49; Fausto, *São Paulo na Primeira República*, 75:14; Lamounier, "Formação de um pensamento político," 359; Levine, *Vale of Tears*, 15; Saes, *Classe média e política*, 56. Quotations from Resende, "O processo político," 99; Dean, "A industrialização durante a República Velha," 277.

49. The rationale for the concessions is in Pará, *Mensagem . . . 1925*, 104–8; Pará, *Mensagem . . . 1928*, 116–18. For examples of oligarchs, see Emmi, *Oligarquia do Tocantins*, 87; Little, *Amazonia*, 36–37; Petit, *Chão de Promessas*, 193–97; Treccani, *Violência e Grilagem*, 106–7.

50. Coimbra, *O Pará na Revolução de 30*, 125–32; quotation from Pará, *Mensagem . . . 1927*, 108.

51. Coimbra, *O Pará na Revolução de 30*, 125–27; Grandin, *Fordlandia*, 9.

52. The bibliography on the new regime is extensive; I rely on Bambirra, "El Estado en Brasil"; Fausto, *A Revolução de 1930*, 61–73; Levine, *Father of the Poor?*, 19–21; Levine, *The Vargas Regime*, 5–9; Skidmore, *Brasil*, 27–32.

53. Levine, *Father of the Poor?*, 25–27, 43; Skidmore, *Brasil*, 55–62.

54. Coimbra, *O Pará na Revolução de 30*, 269–71; Coutinho and Flaksman, "Barata, Magalhães"; Lopes, "A Delegacia Militar do Norte," 9–12; Mesquita, *Magalhães Barata*, 1–14; Rocque, *Magalhães Barata*, 1:155–72.

55. Coutinho and Flaksman, "Barata, Magalhães," 295; Coimbra, *O Pará na Revolução de 30*, 273–308; Meira, *Barata*, 45–47; Rocque, *Magalhães Barata*, 1:174–76, 192–204. Coutinho and Flaksman, "Barata, Magalhães," 295; see also FCTN, "O operariado paraense e suas justas aspirações," *Folha do Norte*, February 10, 1931.

56. FCTN, Cunha Junior, "A Revolução e o povo do interior," *Folha do Norte*, February 28, 1931. See also letter to Clóvis Meira, in Meira, *Barata*, 29; FCTN, "A obra realizada pela Revolução no Pará em contacto com a Metropole e o Sul do Paiz," *Folha do Norte*, March 8, 1931.

57. Decree 184, March 12, 1931, in FCTN, "Para intensificar a agricultura e amparar as classes pobres," *Folha do Norte*, March 13, 1931; "O governo do Estado manda convocar os que desejam lotes gratuitos de terras," *Folha do Norte*, March 31, 1931; "Concessões gratuitas de lotes de terras," *Folha do Norte*, April 2, 1931; "Realizando uma verdadeira correição no baixo Amazonas, o interventor major Barata faz um resumo de seus trabalhos e observações," *Folha do Norte*, November 24, 1931.

58. First quotation from FCTN, "Chave de ouro de um anno de governo amigo da pobreza," *Folha do Norte*, December 25, 1931; see also "A viagem do sr. interventor Federal á região tocantina," *Folha do Norte*, April 14, 1931; quotation on the Panzi case from "Varios despachos do sr. interventor federal do Estado," *Folha do Norte*, January 6, 1931; on the Leite case from "O horario das audiencias do sr. Interventor," *Folha do Norte*, December 25, 1931; see also "Varios despachos do sr. interventor federal do Estado," *Folha do Norte*, January 7, 1931; "Varios despachos do sr. interventor federal do Estado," *Folha do Norte*, January 9, 1931; "Despachos proferidos pelo sr. interventor federal do Estado," *Folha do Norte*, January 13, 1931; "Questão de terras entre posseiros de Monte Alegre e a Companhia Nipponica," *Folha do Norte*, March 8, 1931; Meira, *Barata*, 45; Rocque, *Magalhães Barata*, 1:188.

59. FCTN, "A obra realizada pela Revolução no Pará em contacto com a Metropole e o Sul do Paiz," *Folha do Norte*, March 8, 1931; the phrase "poor people" is from "Realizando uma verdadeira correição no baixo Amazonas, o interventor major Barata faz um resumo de seus trabalhos e observações," *Folha do Norte*, November 24, 1931. See also Levine, *Father of the Poor?*, 59, 150; Ribeiro, "Cartas ao Presidente Vargas," 70.

60. FCTN, "O governo resolveu annular as concessões de terras na zona do Gurupy," *Folha do Norte*, January 16, 1931; "Decreto do sr interventor federal sobre concessões de terras do Estado," *Folha do Norte*, January 16, 1931.

61. FCTN, "Dois decretos despachos do sr. interventor sobre terras pretendidas por menores e concessões declaradas caducas," *Folha do Norte*, January 20, 1931; "As grandes áreas de terras concedidas por opção continuam revertendo ao patrinonio do Estado," *Folha do Norte*, February 25, 1931; "Continuam revertendo ao patrimonio do Estado, grandes áreas de terras, extinctas antigas concessões," *Folha do Norte*, February 25, 1931; "Vao reverter ao patrimonio do Estado mais 2.200.000 hectares de terras," *Folha do Norte*, March 1, 1931; "Foram incorporadas ao patrimonio do Estado mais 2.200.000 hectares de terras, em virtude de terem sido julgadas caducas cinco concessões," *Folha do Norte*, March 3, 1931; "Foi julgada caduca mais uma concessão de 25.000 hectares de terras," *Folha do Norte*, March 6, 1931; "125.000 hectares de terras devolutas, dadas em concessão, que voltam ao patrimonio do Estado," *Folha do Norte*, March 11, 1931; "Concessões de terras annulladas e que são incorporadas ao patrimonio do Estado," *Folha do Norte*, March 17, 1931; "Novas concessões de terras, declaradas caducas, que volvem ao patrimonio do Estado," *Folha do Norte*, April 15, 1931; Rocque, *Magalhães Barata*, 1:197–99.

62. The two quotations are from FCTN, "Continuam revertendo ao patrimonio do Estado, grandes áreas de terras, extinctas antigas concessões," *Folha do Norte*, February 25, 1931, and "Em commissão especial do governo junto á Seretaria de Obras Publicas," *Folha do Norte*, March 13, 1931.

63. FCTN, "As actividades Ford em sua concessão de terras no Tapajós," *Folha do Norte*, March 17, 1931; "Impressões do sr. interventor federal sobre as grandes realizações Ford na Tapajonia," *Folha do Norte*, March 25, 1931.

64. Quotations from FCTN, "A viagem do sr. interventor Federal á região tocantina," *Folha do Norte*, April 14, 1931; the legal decrees are at "Uma solução ao problema da industria extractiva nas terras devolutas do Estado," *Folha do Norte*, January 29, 1931; and "Foi regulada a exploração dos produtos naturaes do Estado," *Folha do Norte*, April 7, 1931; see also FCTN, "Arrendamentos dos castanhaes e balataes de Alemquer," *Folha do Norte*, January 16, 1931; "Reclamações contra esbulhos de terras e violencias," *Folha do Norte*, January 21, 1931; "Não arrendamento de castanhaes municipaes," *Folha do Norte*, November 17, 1931; "A castanha: Perspectivas da nova safra," *Folha do Norte*, December 2, 1931; Borges, *Negociatas Escandalosas*, 44–47; Emmi, *Oligarquia do Tocantins*, 86–88; Rocque, *Magalhães Barata*, 1:188–90.

65. ITERPA, Autos de Medição do Terreno Santa Rosa, José Antonio Picanço Diniz, 1922; ITERPA, *Indice de Títulos Definitivos*, Theodora Gonçalves de Lima, unnamed property (Talonario 932/940, *Provisório* title from June 30, 1921); ITERPA, Autos de Medição "Tres Barracas" ou "São Braz," Theodora Gonçalves de Lima, 1923; see also Emmi, *Oligarquia do Tocantins*, 88.

66. FCTN, "Principaes despachos do sr. interventor," *Folha do Norte*, December 27, 1931; "Solução de queixas dadas directamente ao sr. interventor," *Folha do Norte*, December 13, 1931; quotation from "Em sensacional entrevista á imprensa, o interventor Magalhães Barata narra suas impressões do baixo Amazonas e revela o que pensa sobre os que maldizem de sua pessôa e de sua administração," *Folha do Norte*, December 6, 1931.

67. CF, Autos de Appelação Crime (aka Caso Paiol), *Folha de Obidos*, "O Banditismo nos Castanhaes: Santa Rosa e Seus Sequazes Mattam Barbaramente!," July 3, 1927.

68. CF, Caso Paiol, pp. 495, 519. According to Obidense nun Idaliana Marinho de Azevedo, Santa Rosa was captured and murdered by a military expedition in the early 1930s. Personal communication, May 15, 2009.

69. First quotation from "O Banditismo nos Castanhaes: Sant Rosa e Seus Sequazes Mattam Barbaramente!," *Folha de Óbidos*, July 3, 1927, and second one from "Impressões do sr. Interventor Federal Sobre a Sua Excursão ao Baixo Amazonas," *O Estado do Pará*, December 6, 1931, in CF, Caso Paiol, approx. on p. 449.

70. AMO, Registro de Oficios, 1920–1930, November 11, 1924; Tavares, *Inventário Cultural*, 9, 14–15.

71. Coutinho and Flaksman, "Barata, Magalhães," 295; Fontes, "Cultura e política dos anos trinta," 132–35, 143–45; Levine, *The Vargas Regime*, 48–50; Mesquita, *Magalhães Barata*, 68–95.

72. Emmi, *Oligarquia do Tocantins*, 85–89; Petit, *Chão de Promessas*, 128.

73. ITERPA, Autos de Medição "Tres Barracas" ou "São Braz," Theodora Gonçalves de Lima, 1923; ITERPA, Autos de medição e discriminação, Lote "Central," Fernandes Nunes e Companhia, 1933. See also Borges, *Negociatas Escandalosas*, 67; Emmi, *Oligarquia do Tocantins*, 88–89.

74. Coudreau, *Voyage au Rio Curuá*, 19. This narrative overlapped with another one arguing that the lands of Pacoval were granted by Brazilian emperor Dom Pedro II as a sesmaria or royal land grant. Sesmarias were abolished in the 1820s, but here the Pacovalenses are mixing elements in from their 1877 forced journey to Belém, when one of them was liberated by the emperor. Azevedo, *Puxirum*, 30.

75. Luiz Marques Baptista to Alenquer Mayor, June 27, 1935, in ITERPA, Terras Adjuntas a Felinto, Antônio Vallinoto; APEP, FSP, SCP, Cx 409, Oficios recebidos February 1925–January 1926, Commissioner Manoel Roberto de Azevedo Vasconcellos to the Óbidos Police Chief, December 28, 1925; APEP, FSP, Oficios Recibidos Dezembro 1925–Janeiro 1926, Cx 407, Deodoro Mendonça to State Police Chief, November 30, 1925.

76. APEP, FSP, CP, Cx 434, Oficios recebidos October–November 1926, Alenquer Police Chief to State Police Chief, September 26, 1926.

77. Throughout the decade there were numerous skirmishes and mutinies during the Brazil nut harvest in Alenquer, Obidois, and other municipalities: APEP, FSP, CP, Cx 434, Cx 408, Oficios expedidos December 1925–January 1926, Police Chief to State Secretary General, December 22, 1928; see also letters from December 28, January 6, and especially January 30 showing how police officers used the 1921 report of the Palma Muniz expedition as a blueprint for similar conflicts.

78. Emmi, *Oligarquia do Tocantins*, 90; Mori and Prance, "Taxonomy, Ecology," 141, 144–45.

79. Dezemone, "A Era Vargas e o Mundo Rural Brasileiro"; Petit, *A Esperança Equilibrista*.

Conclusion

1. Furtado, *The Economic Growth of Brazil*, 73, 96.

2. Diacon, *Stringing Together a Nation*; Garfield, *In Search of the Amazon*; Grandin, *Fordlandia*; Hecht, *The Scramble for the Amazon*. Some exceptions are Hemming, *Amazon Frontier*; Neeleman, Neeleman, and Davis, *Tracks in the Amazon*; Stanfield, *Red Rubber, Bleeding Trees*; Weinstein, *The Amazon Rubber Boom*.

3. Hemming, *Amazon Frontier*, 42.

4. Berlin, *Generations of Captivity*, 8–9.

5. Faria, "Identidade e comunidade escrava," 144; Florentino and Góes, *A paz das senzalas*, 124; Klein and Luna, *Slavery in Brazil*, 220–22; Mahony, "Creativity under Constraint," 656–57; Mattos, *Das cores do silêncio*, 52, 110, 141, 216; Metcalf, *Family and Frontier in Colonial Brazil*, 155; Slenes, *Na Senzala, Uma Flor*, 48, 109, 150, 208.

6. First quotation from Marks, *Southern Hunting in Black and White*, 26; the rest are from Stewart, *What Nature Suffers to Groe*, 16.

7. Dean, *With Broadax and Firebrand*, 57; McNeill, "The Ecological Atlantic," 301.

8. For a similar perspective on Brazil as a whole, see Maestri and Fabiani, "O mato." Rice provides an analogous case study: Carney, "Landscapes of Technology Transfer"; Carney and Rosomoff, *In the Shadow of Slavery*; Carney, "'With Grains in Her Hair'"; Hawthorne, *From Africa to Brazil*, 140; Hawthorne, "From 'Black Rice' to 'Brown.'"

9. Heath and Lee, "Memory, Race, and Place," 1360.

10. Cruz, "Puzzling out Slave Origins," 243; De la Torre, "Los ambiguos efectos," 116–17; Fraga Filho, *Encruzilhadas da liberdade*, 192, 290, 300; Gomes, "Roceiros, Mocambeiros," 164–65; Gomes, "Slavery, Black Peasants," 745, 747; Gomes and Machado, "Migraciones, desplazamientos y campesinos negros," 31; Guimarães, *Múltiplos Viveres*, 309, 386; Machado, *O plano e o pânico*, 22, 42–43; Mahony, "Creativity under Constraint," 73; Marques, *Por aí e por muito longe*, 133, 138; Mattos, *Das cores do silêncio*, 361, 376; Paoliello, "Condição camponesa," 235, 242, 246–47; Rios, "My Mother Was a Slave," 116; Rios and Mattos, *Memórias do cativeiro*, 38–39, 50, 221; Rios and Mattos, "Para além das senzalas," 77; Rubert and Silva, "O acamponesamento como sinônimo," 267–70; Weimer, *Os nomes da liberdade*, 30; Yabeta and Gomes, "Memória, cidadania e direitos," 108–9. For broader perspectives beyond Brazil, see Penningroth, "The Claims of Slaves and Ex-Slaves," 1040; Scott and Zeuske, "Demandas de Propiedad y Ciudadanía."

11. Classic examples of this are Reis, *Slave Rebellion in Brazil*, 123, 141; Graham, *Caetana Says No*, 72; Schwartz, *Sugar Plantations*, 245, 252–53; Florentino and Góes, *A paz das senzalas*, 35; Parés, "O Processo de Crioulização"; Sweet, *Recreating Africa*, 34–50.

12. Stewart, *What Nature Suffers to Groe*, 2.

13. Azevedo, *A Brazilian Tenement*, 97–99, 120; Rodrigues, *Os Africanos no Brasil*, 264–65. On Oliveira Vianna, see Dávila, *Diploma of Whiteness*, 22–23; Ramos, "Ciência e racismo," 594–96. Quotation from Cunha, *Os sertões*, 1:1. See also Hermann, "Religião e política," 144–45, 154–55; Levine, *Vale of Tears*, 19–20, 28–29, 47, 91.

14. Two excellent syntheses are Welch et al., *Camponeses brasileiros*, and Motta, "Classic Works." Good examples of this perspective are Guimarães, *Quatro séculos*; Ianni, *A luta pela terra*, 61–89; Palacios, *Cultivadores libres*. The limits of such a perspective are illuminated by French, *Legalizing Identities*, 4; Mattos, "Identidade camponesa," 40–46; Moreira and Hébette, "Metamorfoses de um campesinato," 193, 205; Paoliello, "Condição camponesa," 229, 246–47.

15. Silva and Linhares, *Terra Prometida*; Motta, "Classic Works," 124, 132; Priore and Venâncio, *Uma história*, 45–46; Stédile and Fernandes, *Brava Gente*, 164.

16. See for example Cruz, "Puzzling out Slave Origins"; Ferreira, "A questão racial," 60; Rogers, *The Deepest Wounds*, 131–36.

17. Quotations from French, *Legalizing Identities*, 3; and Ferreira, "O Artigo 68," 8. See also Farfán-Santos, "'Fraudulent' Identities," 129; Kenny, "The Contours of Quilombola Iden-

tity," 141, 154; Leite, "The Trans-Historical," 4–8; Mattos, "Políticas de reparação e identidade coletiva no meio rural," 169, 181.

18. Acevedo Marin, "Quilombolas na Ilha de Marajó," 217. On leaders: Alonso, "O 'movimento' pela identidade," 27; Boyer, "Misnaming Social Conflict"; Carvalho, "O Quilombo da 'Família Silva,'" 37. On language: Acevedo Marin, "Quilombolas na Ilha de Marajó," 217; Carvalho, "O Quilombo da 'Família Silva,'" 42; French, *Legalizing Identities*, 113–15; Moreira and Hébette, "Metamorfoses de um campesinato," 205. On the ties with other communities and agencies: Acevedo Marin and Castro, *Negros do Trombetas*, 32–37; Alonso, "O 'movimento' pela identidade," 32; Bowen, "The Struggle for Black Land Rights in Brazil," 149; Carvalho, "O Quilombo da 'Família Silva,'" 43; French, *Legalizing Identities*, 115–17; Larrea and Ruíz-Peinado, "Memoria y Territorio Quilombola en Brasil," 212–13; Leite, "O Projeto Político Quilombola," 4. On international advocacy groups: Fabiani, "Os Quilombos Contemporâneos," 11. On changes in the collective memory: French, *Legalizing Identities*, 161–70; Malighetti, "Identitarian Politics," 109; Mattos, "Políticas de reparação e identidade coletiva no meio rural," 177–91.

19. French, *Legalizing Identities*, 5, 13–15, 154, 166.

20. Archanjo, "Narrativas de resistência e luta," 6–12; Linhares, "Kilombos of Brazil," 826–31; O'Dwyer, "Os Quilombos," 268–73.

21. McAdam, "Conceptual Origins"; McAdam, *Political Process*, 40–43; Tarrow, *Power in Movement*, 71–90.

Bibliography

Archival Sources

Arquivo do Fórum Judiciário de Santarém
    Livros de Testamentos, 1866–69, 1882–83
    Processos Variados (uncataloged collection)
Arquivo do Museu de Óbidos
    Livros de Sessões do Conselho e da Câmara Municipal, 1840–1930
    Lançamento do Imposto de Industria e Profissão, 1907, 1916
    Registro de Oficios, 1920–30
    Avulsos, 1917–21
Arquivo Público do Estado do Pará
    Documentação Notarial
        Livro de Registro de Imóveis de Óbidos, 1930–33
    Fundo Segurança Pública
        Correspondências dos delegados e subdelegados, 1854–58
        Chefatura de Polícia, 1920–33
    Fundo Segurança Pública, Avulsos
        Chefatura de Polícia, 1919–26
    Juizo de Direito da Primeira Vara
        Inventários Postmortem, 1875–78
    Juizo de Direito da Segunda Vara
        Inventários Postmortem, 1875–77
    Juizo de Órfãos da Capital
        Inventários Postmortem, 1856–69
    Juizo de Órfãos de Igarapé-Miri
        Inventários Postmortem, 1865–76
    Junta Classificadora de Escravos
        Minutas de Ofícios, 1881–85
    Livro de Batismo de Colares, 1895–98
    Repartição de Obras Públicas, Terras e Viação
        Ofícios, 1892–98
    Secretaria do Governo e Secretaria Geral do Estado, Avulsos
        Ofícios para Intendências do Interior, 1920–25
    Secretaria da Presidência da Província do Grão-Pará
        Correspondências das Câmaras Municipais, 1854
Arquivo da Sociedade Literaria e Recreativa 5 de Agosto—Cartório Raiol
    Inventários Postmortem, 1864–86
    Various processes (uncatalogued collection)

Cartório do 2º Ofício ("Ferreira"), Óbidos
  Postmortem Inventories, 1920–36
  Processos Variados, 1912–34
Cartório do 2º Ofício ("Toninho"), Alenquer
  Postmortem Inventories, 1861–93, 1937
Centro de Memória da Amazônia
  Inventários Postmortem, 1871–77, 1929
  14ª Vara Cível, Cartório Sarmento—Autos de Manutenção de Posse, 1862–99,
    1903–13
Fundação Cultural Tancredo Neves
  Newspaper Collection
Instituto de Terras do Pará
  Índice de Títulos Definitivos
  Índice de Títulos de Posse
  Índice de Títulos de Propriedade
  Livro de Títulos de Propriedade 1, Vigia
  Various Processes, Óbidos and Alenquer (selected merchants)
Instituto Histórico e Geográfico do Pará, Arquivo Palma Muniz
  Câmara de Municipal da Vila de Alenquer, Livro de Atas 1855–62
Paróquia de Alenquer
  Livro de Baptismo, 1895–99, 1910–11
  Livro de Baptismo, 1935–37
  Livro de Casamentos, 1891–1901
  Livro de Óbitos, 1918–21
Sterling Memorial Library—Yale University
  United States Department of State. Dispatches from U.S. consuls in Pará, Brazil,
    1831–1906

Interviews

ALENQUER

António Nâcio Vianna (Tio Nácio, born 1921), Pacoval (conducted in Alenquer), May 8,
  2009
Dona Cruzinha (Maria da Cruz, born 1944), Pacoval, April 20, April 21, April 23, 2009
Dona Mimita (Lucimar, born 1920), Pacoval, April 23, 2009
Dona Nazita (born 1946), Pacoval, April 29, 2009
Dona Piquixita (Maria José Monteiro, born 1915), Pacoval, April 22, 2009
Olinda Vallinoto (born c. 1916), Alenquer, April 20, 2009

ORIXIMINÁ

Anarcindo da Silva Cordeiro (born 1951), Tapagem community (conducted in
  Oriximiná), June 7, 2009
Antônio Souza (born 1940) and Edith Printes (born 1943), Abuí community, June 5 and
  June 6, 2009
Deometilo Cordeiro (born 1945), Tapagem community, June 6, 2009

Dona Biquinha (born 1934), Pancada community (conducted in Oriximiná), May 29, 2009

Dona Deuza (born 1954), Pancada community (conducted in Oriximiná), May 29, 2009

Francisco Alegre (born 1952), Boa Vista, June 4, 2009

Francisco Edilberto Figueiredo de Oliveira (born 1958), Poço Fundo (conducted in Oriximiná), May 29, 2009

João Xavier (born 1952), Abuí (conducted in the Tapagem community), June 4, 2009

José do Carmo (born 1944), Jamary, May 27, 2009

José Melo (born 1942), Boa Vista, May 27, 2009

Luis Bacellar Guerreiro (born 1929), June 9, 2009

Maria de Souza (born 1935), Javary (conducted in Oriximiná), May 23, 2009

Manoel das Graças Pereira (born 1951), Nova Esperança, Erepecú (conducted in Boa Vista), May 25, 2009

Manoel Francisco Cordeiro Xavier (born 1934), Abuí community, June 5 and June 6, 2009

Maria Rosa Xavier Cordeiro (born 1925), Tapagem community, June 5 and June 6, 2009

Group interview with Nicanor (born 1940), Aldenor Pereira de Jesus (born 1953), Teresa Fernandes Regis (born 1938), and Raymundo Dias Barbosa (born 1947), all from Erepecú Lake (conducted in Jamary), June 6, 2009

Ruy Brasil (born 1945), Tapagem community (conducted in Oriximiná), June 7, 2009

Valério and Zuleide dos Santos (born 1945 and 1955), Boa Vista, May 26, 2009

SANTARÉM

Cleinilze Souza Silva (born 1954), Santarém, April 27, 2009

VIGIA

Alcides Souza de Jesus (born 1940), Vigia, March 26, 2009

Ana Maria dos Santos (unknown DOB), Vigia, August 17, 2009

Avelino de Almeida (unknown DOB), Cacau (conducted in Vigia), March 10, 2009

Dona Osmarina de Melo (unknown DOB, c. 1948), Cacau, March 10, 2009

Guilhermina da Conceição Goulart (born 1916), Cacau (conducted in Vigia), March 13, 2009

Manoel Ramos dos Santos (born 1948), Terra Amarela, March 19, 2009

Manoel Santana Ferreira (born 1940), Vigia, March 25, 2009

Manoel Santana Porto de Miranda (born 1940), Vigia, March 9, 2009

Nadi Ferreira dos Santos (born 1953), Terra Amarela, March 19, 2009

Raimunda das Dores Miranda (Dona Bena, born 1922), Santo Antônio do Tauapará, April 1, 2009

Seu Cebola (Ilson Pereira de Melo, born 1940), Cacau (conducted also in Terra Amarela and Vigia), March 10, March 19, August 12, 2009

Seu Diquinho (unknown DOB), Cacau, March 10, 2009

Seu Nunhes (Manoel da Conceição de Mello, born 1926), Vigia, March 11, March 31, August 10, 2010

Sylvia Helena Tocantins de Mello Éder (born 1933), Belém, March 3, 2009

Zacarias Atayde (born 1943), Santo Antônio de Tauapará, April 1, 2009

Newspapers and Periodicals

A Provincia do Pará. Belém, 1921.
Diário Oficial do Estado do Pará. Belém, 1921.
Folha do Norte. Belém, 1921–1935.
O Alenquerense. Alenquer, 1942.
O Município. Alenquer, 1932.

Published Primary Sources

Agassiz, Louis, and Elizabeth Cabot Cary Agassiz. *A Journey in Brazil*. Boston, MA:
    Houghton Mifflin, 1895.
Aliprandi, Ermenegildo, and Virginio Martini. *Gli Italiani del Nord del Brasile: Rassegna
    delle Vite e delle Opere della Stirpe Italica negli Stati del Nord Brasiliano*. Belém: Typ. da
    Livraria Gillet, 1932.
Amaral, Antonio. *Memorias para a História da Vida do Venerável Arcebispo de Braga Fr.
    Caetano Brandão*. Vol. 1. Braga: Typ. dos Orfãos, 1867.
Azevedo, Aluísio. *A Brazilian Tenement*. New York: R. M. McBride, 1926.
Bastos, A. C. Tavares. *O Vale do Amazonas: A livre navegação do Amazonas, estatística,
    produção, comércio, questões fiscais do Vale do Amazonas*. São Paulo: Editôra Nacional,
    1975.
Bates, Henry Walter. *The Naturalist on the River Amazons*. London: John Murray, 1892.
Belmar, Alejandro de. *Voyage aux Provinces Brésiliennes du Pará et des Amazones en 1860,
    Précédé d'une Rapide Coup d'Oeil, sur le Littoral du Brésil*. London: Trezise Imprimeur,
    1861.
Bentley, William Holman. *Life on the Kongo*. Oakland, CA: Pacific Press, 1891.
Biard, François Auguste. *Deux Années au Brésil*. Paris: L. Hachette, 1862.
Borges, Antonio. *Negociatas Escandalosas*. Rio de Janeiro: Typographia do Jornal de
    Commercio, 1938.
Brazil. *Recenseamento do Brasil em 1872*. 12 vols. Rio de Janeiro: Typ. G. Leuzinger, 1874.
———. *Recenseamento do Brasil em 1872: Pará*. Vol. 5. Rio de Janeiro: Typ. G. Leuzinger,
    1874.
Brown, Charles Barrington, and William Lidstone. *Fifteen Thousand Miles on the Amazon
    and Its Tributaries*. London: E. Stanford, 1878.
Capelo, Hermenegildo, and Roberto Ivens. *From Benguella to the Territory of Yacca:
    Description of a Journey into Central and West Africa. Comprising Narratives, Adventures,
    and Important Surveys of the Sources of the Rivers, Cunene, Cubango, Luando, Cuanza, and
    Cuango, and of Great Part of the Course of the Two Latter; Together with the Discovery of
    the Rivers Hamba, Cauali, Sussa, and Cugho, and a Detailed Account of the Territories of
    Quiteca, N'bungo, Sosso, Futa, and Yacca*. London: S. Low, Marston, Searle, & Rivington,
    1882.
Champney, Elizabeth Williams. *Three Vassar Girls in South America: A Holiday Trip of
    Three College Girls through the Southern Continent, up the Amazon, down the Madeira,
    across the Andes, and up the Pacific Coast to Panama*. Boston: Estes and Lauriat
    Publishers, 1885.

Cordeiro, Luiz. *O Estado do Pará: Seu commercio e industrias de 1719 a 1920.* Belém: Tavares Cardoso & Ca., 1920.

Coudreau, Otille. *Voyage à la Mapuerá: 21 avril 1901–24 décembre, 1901.* Paris: A. Lahure, 1903.

———. *Voyage au Cuminá: 20 Avril 1900–7 Septembre 1900.* Paris: Lahure, 1901.

———. *Voyage au Rio Curuá: 20 Novembre 1900–7 Mars 1901.* Paris: Lahure, 1903.

Coudreau, Otille, and Henri Coudreau. *Voyage au Trombetas: 7 Août 1899–25 Novembre 1899.* Paris: Lahure, 1900.

Cruls, Gastão. *A Amazônia que eu vi: Óbidos—Tumucumaque.* Rio de Janeiro, 1930.

Cunha, Euclides da. *Os sertões.* Vol. 1. 2 vols. Rio de Janeiro: Ministério da Cultura / Fundação Biblioteca Nacional / Departamento Nacional do Livro, Unknown date.

Derby, Orville. "O Rio Trombetas." *Boletim do Museu Paraense de Historia Natural e Ethnographia* 2 (1898): 366–82.

Du Chaillu, Paul. *Voyages et aventures dans l'Afrique équatoriale moeurs et coutumes des habitants . . . par Paul du Chaillu.* Paris: M. Lévy Frères, 1863.

Ducke, Adolpho. "Explorações Scientíficas no Estado do Pará." *Boletim do Museu Paraense Emilio Goeldi* 7 (1909).

Edwards, William H. *A Voyage up the River Amazon Including a Residence at Pará.* London: John Murray, Albemarle Street, 1861.

Frazão, Carlos. *Castanha do Brasil (Brazil Nuts): Quatro castanhas equivalem a dois ovos de gallinha.* Belém: Unknown publisher, 1935.

Frias, David Correia Sanches de. *Uma viagem ao Amazonas.* Lisboa: Typ. de Mattos Moreira & Cardosos, 1883.

Great Britain Foreign Office. *Reports from Her Majesty's Consuls on the Manufactures, Commerce, &c. of Their Consular Districts.* London: Printed by Harrison and Sons, 1864.

Hartt, Charles F. "A Geologia do Pará." *Boletim do Museu Paraense de Historia Natural e Ethnographia* I (1896): 257–73.

Herndon, Lieut. Wm. Lewis. *Exploration of the Valley of the Amazon.* Washington, DC: Taylor & Maury, 1854.

Hilbert, Peter Paul. *A cerâmica arqueológica da região de Oriximiná.* Belém: Instituto de Antropologia e Etnologia do Pará, 1955.

Instituto Brasileiro de Geografia e Estatistica. "Alenquer—PA." Edited by Instituto Brasileiro de Geografia e Estatistica. *Enciclopédia dos Municípios Brasileiros.* Rio de Janeiro, 1957.

———. *Recenseamento Geral do Brasil (Realizado em 1 de Setembro de 1940): Série Regional Parte III—Pará: Censo Demográfico População e Habitação: Censos Econômicos Agrícola, Industrial, Commercial e de Serviços.* Rio de Janeiro: Serviço Gráfico do IBGE, 1952.

Instituto Historico e Geographico do Pará. *Catálogo da primera série de uma galeria histórica.* Belém: Imprensa Official do Estado do Pará, 1918.

Kidder, Daniel P. *Sketches of Residence and Travels in Brazil, Embracing Historical and Geographical Notices of the Empire and Its Several Provinces.* Vol. 2. London: Wiley and Putnam, 1845.

Kingston, William Henry Giles. *On the Banks of the Amazon or a Boy's Journal of His Adventures in the Tropical Wilds of South America.* London: T. Nelson and Sons, 1872.

Lage, Sandoval. *Quadros da Amazônia: Pref. de Victor do Espíritu Santo*. Rio de Janeiro: Oficinas Gráficas Espirito Santo, 1944.

Le Cointe, Paul. *Amazônia brasileira III: Arvores e plantas úteis (indígenas e aclimadas): Nomes vernáculos e nomes vulgares. Classificação botânica. Habitat. Principais aplicações e propriedades*. São Paulo: Companhia Editora Nacional, 1934.

———. *L'Amazonie Bresilienne: Le Pays-Ses Habitants, Ses Ressources, Notes et Statistiques Jusqu'en 1920*. Vol. 2. Paris: Augustin Challamel, 1934.

Lima, Joaquim. "História dos Negros que Através da Luta Conseguiram Libertar-se dos Senhores de Escravos de Santarém, Pará." Mimeograph. Oriximiná, 1992.

Lopes, F. Gonçalves. *Emilio Carrey: O Amazonas: Segunda parte: Os revoltosos do Pará: Descripção de viagem, traduzida e annotada por F.F. da Silva Vieira*. Lisboa: Typographia do Futuro, 1862.

Magalhães, José Vieira Couto de. *Relatório dos negocios da Provincia do Pará*. Belém: Typographia de Frederico Rhossard, 1864.

Maury, Matthew Fontaine. *The Amazon, and the Atlantic Slopes of South America: A Series of Letters Published in the National Intelligencer and Union Newspapers, under the Signature of "Inca."* Washington, DC: F. Taylor, 1858.

Maw, Henry Lister. *Journal of a Passage from the Pacific to the Atlantic Crossing the Andes in the Northern Provinces of Peru, and Descending the River Marañon, or Amazon*. London: John Murray, 1829.

Mesquita, Lindolfo. *Magalhães Barata, o Pará e sua história*. Belém: Jornal do Commercio, 1944.

Miranda, Vicente Chermont de. *Glossario paraense: Ou, Collecção de vocabulos peculiares á Amazonia e especialmente á ilha do Marajó, por Vicente Chermont de Miranda*. Belém: Livraria Maranhense, 1906.

Monteiro, Joachim John. *Angola and the River Congo*. Vol. 1. London: Macmillan and Company, 1875.

Moraes, Antonio José de Mello. *Corographia Historica, Chronographica, Genealogica, Nobiliaria, e Politica do Imperio do Brasil*. Vol. II. Rio de Janeiro: Typographia Brasileira, 1860.

Morais, Joaquim de Almeida Leite. *Apontamentos de viagem*. Unknown publisher, 1882.

Muniz, João de Palma. "Adhesão do Grão-Pará à Independência." *Revista do Instituto Histórico e Geográfico do Pará* 6 (1922).

———. *Castanhaes de Alemquer: Relatório de verificação local, apresentado ao Sr. Dr. Antonino de Souza Castro, Governador do Estado pelo engenheiro civil João de Palma Muniz, Chefe da 3ª Secção da Directoria de Obras Publicas*. Belém, 1921.

———. *Estado de Grao-Pará: Inmigração e Cólonisação: História e Estatistica*. Belém: Imprensa Official do estado do Pará, 1916.

———. "Grenfell na história do Pará, 1823–1824." *Annaes da Biblioteca e Archivo Público* 10 (1926).

———. *Indice Geral dos Registros de Terras*. Vol. 1. Belém: Imprensa Official, 1907.

———. *Legislação de Terras: Dados Estadísticos*. Belém: Departamento de Obras Públicas, Terras e Viação, 1924.

———. *Limites Pará-Goyaz: Notas e Documentos*. Belém: Imprensa Official do Estado, 1920.

———. *O Instituto Santo Antonio do Prata (Município de Igarapé-Assú)*. Belém: Typ. da Livraria Escolar, 1913.

———. "Os Contemplados (Não Contemplados com Documentação)." *Revista do Instituto Historico e Geographico do Pará* 1 (1912): 71–78.

———. *Patrimonios dos Conselhos Municipaes do Estado do Pará*. Paris: Aillaud & Cia., 1904.

Nicholson, John. *The Operative Mechanic and British Machinist; Being a Practical Display of the Manufactures and Mechanical Arts of the United Kingdom*. London: Knight and Lacey, Paternoster-Row; and Westley and Tyrrell, 1825.

Orton, James. *The Andes and the Amazon, or Across the Continent of South America*. New York: Harper & Brothers Publishers, 1870.

Pará. *Discurso Recitado pelo Exm.o Sñr Doutor João Maria de Moraes, Vice-prezidente da Provincia do Pará na Abertura da Segunda Sessão da Quarta Legislatura da Assembléa Provincial no dia 15 de Agosto de 1845*. Belém: Typ. de Santos & Filhos, 1845.

———. *Exposição Apresentada pelo Exm.o Senr. Conselheiro Sebastião do Rego Barros, Presidente da Provincia do Gram-Pará, ao Exm.o Senr Tenente Coronel d'Engenheiros Henrique de Beaurepaire Rohan, no Dia 29 de Maio de 1856, por Occasião de passar-lhe a Administração da Mesma Provincia*. Belém: Typ. de Santos e Filhos, 1856.

———. *Falla Dirigida pelo Exmo. Sr. Conselheiro Jeronimo Francisco Coelho, Prezidente da Provincia do Gram Pará a Assembléa Legislativa Provincial na Abertura da Segunda Sessão Ordinaria da Sexta Legislatura no Dia 1º de Outubro de 1848*. Belém: Typographia de Santos & Filhos, 1848.

———. *Falla dirigida pelo exm.o sñr conselheiro Jeronimo Francisco Coelho, prezidente da provincia do Gram Pará á Assembléa Legislativa Provincial na abertura da segunda sessão ordinaria da sexta legislatura no dia 1.o de outubro de 1849*. Belém: Typ. de Santos & filhos, 1849.

———. *Mensagem Apresentada ao Congresso Legislativo do Estado do Pará em sessão solemne de abertura da 1ª reunião de sua 10ª legislatura a 7 de Setembro de 1918*. Belém: Typ. da Imprensa Official do Estado, 1918.

———. *Mensagem Apresentada ao Congresso Legislativo do Estado do Pará em sessão solemne de abertura da 2ª reunião de sua 10ª legislatura a 7 de Setembro de 1919 pelo Governador do Estado Dr. Lauro Sodré*. Belém: Typ. da Imprensa Official do Estado, 1919.

———. *Mensagem Apresentada ao Congresso Legislativo do Estado em Sessão Solenne de Apertura da 2ª Reunião de sua 12ª Legislatura, a 7 de Setembro de 1925, pelo Gobernador do Estado Dr. Dionysio Ausier Bentes*. Belém: Officinas Graphicas do Instituto Lauro Sodré, 1925.

———. *Mensagem Apresentada ao Congresso Legislativo do Estado, em Sessão Solenne de Abertura da 1ª Reunião de sua 13ª Legislatura, a 7 de Setembro de 1927, pelo Governador do Estado Dr. Dionysio Ausier Bentes*. Belém: Officinas Graphicas do Instituto Lauro Sodré, 1927.

———. *Mensagem Apresentada ao Congresso Legislativo do Estado, em sessão solenne de abertura da 2ª reunião de sua 13ª Legislatura, a 7 de Setembro de 1928, pelo Governador do Estado, Dr. Dionysio Ausier Bentes*. Belém: Officinas Graphicas do Instituto Lauro Sodré, 1928.

———. *Mensagem Dirigida Apresentada ao Congresso Legislativo do Estado do Pará em Sessão Solemne de Abertura da 3ª Reunião de sua 10ª Legislatura a 7 de Setembro de 1920 pelo Governador do Estado Dr. Lauro Sodré*. Belém: Typographia da Imprensa Official do Estado, 1920.

———. *Relatorio 1855, em 15 de Outubro*. Belém, 1855.

———. *Relatorio Apresentado á Assembléa Legislativa da Provincia do Pará na Primeira Sessão da XIII Legislatura Pelo Exm.o Senr. Presidente da Provincia, Dr. Francisco Carlos de Araujo Brusque em 1.o de Setembro de 1862*. Belém: Typ. de Frederico Carlos Rhossard, 1862.

———. *Relatório Apresentado á Assembléa Legislativa Provincial na 2.a Sessão da 22.a Legislatura em 15 de Fevereiro de 1881 pelo Exm. Sr. Dr. José Coelho da Gama e Abreu*. Belém: Typ. do Diario de Noticias de Costa & Campbell, 1881.

———. *Relatorio Apresentado á Assemblea Legislativa Provincial por S. Exca. o Sr. Vice-Almirante e Conselheiro de Guerra Joaquim Raymundo de Lamare, Presidente da Provincia, em 15 de Agosto de 1867*. Belém: Typ. de Frederico Rhossard, 1867.

———. *Relatorio Apresentado ao Exm. Senr. Dr. Francisco Maria Corrêa de Sá e Benevides pelo Exm. Senr. Dr. Pedro Vicente de Azevedo, por Occasião de Passar-lhe a Administração da Provincia do Pará, no Dia 17 de Janeiro de 1875*. Belém: Typ. de F.C. Rhossard, 1875.

———. *Relatorio apresentado ao exm.o snr. dr. José Joaquim da Cunha, presidente da provincia do Gram Pará, pelo commendador Fausto Augusto 'Aguiar por occasião de entregar-lhe a administração da provincia no dia 20 de agosto de 1852*. Belém: Typ. de Santos & Filhos, 1852.

———. *Relatorio do Presidente da Provincia do Gram Pará o Exmo. Snr. Dr. Fausto Augusto d'Aguiar na Abertura da Segunda Sessão Ordinaria da Setima Legislatura da Assemblea Provincial No dia 15 de Agosto de 1851*. Belém: Typ. de Santos & Filhos, 1851.

———. *Relatorio Lido pelo Ex.mo S.r Vice-presidente da Provincia, d.r Ambrosio Leitão da Cunha, na Abertura da Primeira Sessão Ordinaria da XI. Legislatura da Assemblea Legislativa Provincial no dia 15 de Agosto de 1858*. Belém: Typ. Commercial de Antonio José Rabello Guimarães, 1858.

Penna, Domingos Soares Ferreira. *A região Occidental da Provincia do Pará. Resenhas estatisticas das comarcas de Obidos e Santarem apresentadas a S. Ecx. o Sr. Conselheiro José Bento da Cunha Figueiredo Presidente da Provincia*. Belém: Typographia do Diario de Belem, 1869.

———. *Noticia Geral das Comarcas de Gurupá e Macapá*. Belém: Typographia do Diario do Gram-Pará, 1872.

———. *Obras Completas de Domingos Soares Ferreira Penna*. Vol. II. Belém: Conselho Estadual de Cultura, 1971.

Rodrigues, João Barbosa. *Exploração e Estudo do Vale do Amazonas: O Rio Capim*. Rio de Janeiro: Typ. de G. Leuzinger & Filhos, 1890.

———. *Exploração e Estudo do valle do Amazonas: Rio Tapajós*. Rio de Janeiro: Typographia Nacional, 1875.

———. *Exploração e Estudo do Valle do Amazonas: Rio Trombetas*. Rio de Janeiro: Typographia Nacional, 1875.

———. *Poranduba amazonense: ou, Kochiymauara porandub*. Rio de Janeiro: G. Leuzinger & filhos, 1890.

Secção de Estatistica e Terras da Prefeitura Municipal de Alemquer. *Apontamentos sobre o Município de Alemquer collogidos para a Primeira Feira de Amostras do Pará*. Belém: Typographia da União Espirita Paraense, 1937.

Simões, Fulgêncio. *Municipio de Alemquér: Seu Desenvolvimento Moral e Material e Seu Futuro*. Belém: Typographia da Livraria Loyola, 1908.

Skinner, John S. "Tide Mills of Easton, MD." In *The American Farmer: Containing Original Essays and Selections on Agriculture, Horticulture, Rural and Domestic Economy, and Internal Improvements: With Illustrative Engravings and the Prices of Country Produce*, edited by John S. Skinner, X:29. Baltimore, MD: John D. Toy, 1828.

Smith, Herbert Huntington. *Brazil, the Amazons and the Coast*. London: S. Low, Marston, Searle, and Rivington, 1879.

Smyth, William, and Frederick Lowe. *Narrative of a Journey from Lima to Pará across the Andes and down the Amazon*. London: John Murray, Albermarle Street, 1836.

Sousa, Nicolino José Rodrigues de. *Diário das Três Viagens (1877–1878–1882) do Revmo. Padre Nicolino José Rodrigues de Sousa ao Rio Cuminá*. Rio de Janeiro: Imprensa Nacional, 1946.

Souza, Inglês de. *O Cacaulista (Cenas da Vida do Amazonas)*. Belém: Universidade Federal do Pará, 1973.

———. *O Coronel Sangrado (Cenas da Vida no Amazonas)*. Belém: UFPA, 1968.

Spix, Johann Baptist von, and Karl Friedrich Philipp von Martius. *Reise in Brasilien: auf Befehl Sr. Majestät Maximilian Joseph I., Königs von Baiern, in den Jahren 1817 bis 1820 gemacht und beschrieben*. Munich: M. Lindauer, 1823.

Spruce, Richard. *Notes of a Botanist on the Amazon & Andes*. Edited by Alfred Russel Wallace. London: Macmillan and Co., 1908.

Veríssimo, José. "Ethnographia." In *O Pará em 1900*, 135–80. Belém: Imprensa de Alfredo Augusto Silva, 1900.

———. *José Verissimo: A Pesca na Amazônia*. Vol. 111. Monographias Brasileiras. Rio de Janeiro: Livraria Classica de Alves & C., 1895.

Wallace, Alfred Russel. *Travels on the Amazon and Rio Negro*. London: Ward, Lock, and Co., 1889.

Warren, John Esaias. *Para; or Scenes and Adventures on the Banks of the Amazon*. New York: G. P. Putnam, 1851.

Weeks, John H. *Among the Primitive Bakongo: A Record of Thirty Years' Close Intercourse with the Bakongo and Other Tribes of Equatorial Africa*. London: Seeley, Service, and Co., 1914.

Secondary Sources

Abrams, Leonard. *Quilombo Country*. New York: Moving Eye Productions, 2006.

Acevedo Marin, Rosa Elizabeth. "Alianças Matrimoniais na Alta Sociedade Paraense no Século XIX." *Estudos Econômicos* 15 (1985): 153–67.

———. *Julgados da Terra: Cadeia de Apropriação e Atores Sociais em conflito na Ilha de Colares*. Belém: UFPA, 2004.

———. "Quilombolas na Ilha de Marajó: Território e organização política." In *Diversidade do campesinato: expressões e categorias*, 1. Construções identitárias e sociabilidades:

209–27. São Paulo: Editora UNESP; Núcleo de Estudos Agrários e Desenvolvimento Rural, 2009.

Acevedo Marin, Rosa Elizabeth, and Edna Castro. *Negros do Trombetas: Guardiães de Matas e Rios.* Belém: CEJUP-UFPA, 1998.

———. *No caminho das pedras de Abacatal: Experiência social de grupos negros no Pará.* Belém: UFPA / NAEA, 2004.

Agorsah, Emmanuel Kofi. *Maroon Heritage: Archaeological, Ethnographic, and Historical Perspectives.* Barbados: Canoe Press, 1994.

Alberto, Paulina. *Terms of Inclusion: Black Intellectuals in Twentieth-Century Brazil.* Chapel Hill: University of North Carolina Press, 2011.

Alden, Dauril. "The Significance of Cacao Production in the Amazon Region during the Late Colonial Period: An Essay in Comparative Economic History." *Proceedings of the American Philosophical Society* 120 (1976): 103–35.

Almeida, Alfredo Wagner Berno de. *Tierras tradicionalmente ocupadas.* Buenos Aires: Teseo, 2009.

Almeida, José Jonas. "Do extrativismo à domesticação: as possibilidades da castanha-do-pará." PhD diss., University of São Paulo, 2015.

Almeida, Wilkler. *Tauapará.* Vigia, PA: Edição do Autor, 2005.

Alonso, Sara. "O 'movimento' pela identidade e 'resgate das terras de preto': uma prática de socialização." In *Prêmio ABA/MDA Territórios Quilombolas,* 13–36. Brasília: Ministério do Desenvolvimento Agrário / Núcleo de Estudos Agrários e Desenvolvimento Rural, 2006.

Amaral, Assunção José Pureza, and Raimundo Paulo Cordeiro. "Entre homens e mulheres, escravizados e libertos, campo e cidade—eis as tias 'negras' do carimbó na fronteira do saber na cidade da Vigia-PA." *Revista Cadernos do Ceom* 25, no. 37 (2012): 139–60.

Anderson, Robin. "Following Curupira: Colonization and Migration in Pará, 1758 to 1930 as a Study in Settlement of the Humid Tropics." Unpublished PhD diss., University of California-Davis, 1976.

Anderson, Scott Douglas. "Sugarcane on the Floodplain: A Systems Approach to the Study of Change in Traditional Amazonia." PhD diss., University of Chicago, 1993.

Andrade, Lúcia Mendonça Morato de. "Os Quilombos da Bacia do Rio Trombetas: Breve Histórico." *Revista de Antropologia* 38, no. 1 (1995): 79–99.

———. *Quilombola Lands in Oriximiná: Pressure and Threats.* São Paulo: Comissão Pró-Índio de São Paulo, 2011.

Andrews, George Reid. *Afro-Latin America: 1800–2000.* New York: Oxford University Press, 2004.

———. *Blackness in the White Nation: A History of Afro-Uruguay.* Chapel Hill: University of North Carolina Press, 2010.

Ângelo, Helder Bruno Palheta. "A Trajetória dos Corrêa de Miranda no Século XIX: Alianças Sociais, Base Econômica e Capital Simbólico." MA thesis, Universidade Federal do Pará, 2009.

Archanjo, Elaine Cristina Oliveira Farias. "Narrativas de resistência e luta pela terra de quilombolas de Boa Vista, Município de Oriximiná-PA," 1–13. Parintins, 2015.

Armstrong, Douglas V., and Elizabeth Jean Reitz. *The Old Village and the Great House: An Archaeological and Historical Examination of Drax Hall Plantation, St. Ann's Bay, Jamaica.* Chicago: University of Illinois Press, 1990.

Arregui, Aníbal. "Amazonian Quilombolas and the Technopolitics of Aluminum." *Journal of Material Culture* 20, no. 3 (2015): 249–72.

Arruti, José Maurício. "Comunidades remanescentes de quilombos." *Tempo e Presença,* 2000, 25–28.

Associação Brasileira de Antropologia, ed. *Prêmio ABA/MDA Territórios Quilombolas.* Brasília: Ministério do Desenvolvimento Agrário / Núcleo de Estudos Agrários e Desenvolvimento Rural, 2006.

Assunção, Matthias Röhrig. "A memória do tempo de cativeiro no Maranhão." *Tempo* 15, no. 29 (2010): 67–110.

———. *Capoeira: The History of an Afro-Brazilian Martial Art.* London: Routledge, 2005.

Azevedo, Esterzilda de. *Arquitetura do açúcar: Engenhos do Recôncavo Baiano no Período Colonial.* São Paulo: Nobel, 1990.

Azevedo, Idaliana Marinho de. *Puxirum: Memória dos Negros do Oeste Paraense.* Belém: Instituto de Artes do Pará, 2002.

Baker, Alan R. H. "Introduction: On Ideology and Landscape." In *Ideology and Landscape in Historical Perspective: Essays on the Meanings of Some Places in the Past,* 1–14. Cambridge: Cambridge University Press, 1992.

Bambirra, Vania. "El Estado en Brasil: Del Dominio Oligárquico a la Apertura Controlada." In *El Estado en América Latina: Teoría y Práctica,* edited by Pablo González Casanova, 247–66. México: Siglo XXI, 1990.

Barbara, Vanessa. "In Denial over Racism in Brazil." *New York Times,* March 23, 2015.

Barcia, Manuel. *Seeds of Insurrection: Domination and Resistance on Western Cuban Plantations, 1808–1848.* Baton Rouge: Louisiana State University Press, 2008.

Barickman, Bert J. "'A Bit of Land, which They Call Roça': Slave Provision Grounds in the Bahian Recôncavo, 1780–1860." *Hispanic American Historical Review* 74 (1994): 649–87.

Baron, Robert. "Amalgams and Mosaics, Syncretisms and Reinterpretations: Reading Herskovits and Contemporary Creolists for Metaphors of Creolization." *Journal of American Folklore* 116, no. 459 (2003): 88–115.

Barroso, Daniel Souza. "Coletando o cacau 'bravo', plantando o cacau 'manso' e outros gêneros: um estudo sobre a estrutura da posse de cativos no Baixo Tocantins (Grão-Pará, 1810–1850)." São Paulo, 2014.

———. "Múltiplos do Cativeiro: Casamento, compadrio e experiência comunitária numa propriedade escrava no Grão-Pará (1840–1870)." *Afro-Ásia,* no. 50 (2014): 93–128.

Batista, Luciana Marinho. "Demografia, família e resistência no Grão-Pará (1850–1855)." In *Terra Matura: Historiogafía e História Social na Amazônia,* edited by José Maia Bezerra Neto and Décio de Alencar Guzmán, 207–41. Belém: Paka-Tatu, 2002.

———. "Muito Além dos Seringais: Elites, Fortunas e Hierarquias no Grão-Pará, c. 1850–c. 1870." Unpublished PhD diss., Universidade Federal do Rio de Janeiro, 2004.

Beatty-Medina, Charles. "Between the Cross and the Sword: Religious Conquest and Maroon Legitimacy in Sixteenth and Early Seventeenth Century Esmeraldas." In

Africans to Spanish America: Expanding the Diaspora, 95–113. Urbana: University of Illinois Press, 2012.

Bergad, Laird W. "Demographic Change in a Post-Export Boom Society: The Population of Minas Gerais, Brazil, 1776–1821." *Journal of Social History* 29, no. 4 (1996): 895–932.

Berlin, Ira. *Generations of Captivity: A History of African-American Slaves*. Cambridge, MA: Belknap Press of Harvard University Press, 2003.

Berlin, Ira, and Philip D. Morgan. "Introduction." In *The Slaves' Economy: Independent Production by Slaves in the Americas*, edited by Ira Berlin and Philip D. Morgan, 1–27. London: Frank Cass, 1991.

———. *The Slaves' Economy: Independent Production by Slaves in the Americas*. London: Frank Cass, 1995.

Besson, Jean. "The Appropriation of Lands of Law by Lands of Myth in the Caribbean Region." In *Land, Law and Environment: Mythical Land, Legal Boundaries*, 116–35. London: Pluto Books, 2000.

Bezerra Neto, José Maia. "Escravidão e crescimento econômico no Pará (1850–1888)," 1–16, 2008.

———. *Escravidão negra no Grão-Pará (séculos XVII-XIX)*. 2nd ed. Belém: Editora Paká-Tatu, 2012.

———. "Histórias Urbanas de Liberdade: Escravos em Fuga na Cidade de Belém, 1860–1888." *Afro-Ásia* 28 (2002): 221–50.

———. "Os fundadores de 1917, herdeiros de 1900? IHGP 90 Anos: História, Memória e Tradições." IHGP, 2007.

———. "Por todos os meios legítimos e legais: As lutas contra a Escravidão e os limites da Abolição (Brasil, Grão-Pará: 1850–1888)." Unpublished PhD diss., Pontifícia Universidade Católica de São Paulo, 2009.

Bieber, Judy. *Power, Patronage, and Political Violence: State Building on a Brazilian Frontier, 1822–1889*. Lincoln: University of Nebraska Press, 1999.

Bilby, Kenneth. *True-Born Maroons*. Gainesville: University Press of Florida, 2008.

Blomley, Nicholas. "Landscapes of Property." *Law & Society Review* 32 (1998): 567–612.

Bolland, O. Nigel. "The Politics of Freedom in the British Caribbean." In *Struggles for Freedom: Essays on Slavery, Colonialism, and Culture in the Caribbean and Central America*, 163–91. Belize: Angelus Press, 1997.

———. "'Proto-Proletarians?' Slave Wages in the Americas: Between Slave Labour and Free Labour." In *Struggles for Freedom: Essays on Slavery, Colonialism, and Culture in the Caribbean and Central America*, 101–31. Belize: Angelus Press, 1997.

Bondar, Gregório. *A cultura do cacau na Bahia*. Instituto de Cacau da Bahia, 1938.

Borges, Ricardo. *Castanha e Oleaginosas da Amazônia*. Belém: Associação Comercial do Pará, 1952.

———. *Vultos Notáveis do Pará*. Belém: CEJUP, 1986.

Bowater, Donna. "Brazil's Quilombos, Founded by Escaped Slaves, Offer a Window to the Past." *Al Jazeera America*, October 12, 2014.

Bowen, Merle L. "The Struggle for Black Land Rights in Brazil: An Insider's View on Quilombos and the Quilombo Land Movement." *African and Black Diaspora: An International Journal* 3, no. 2 (2010): 147–68.

Boyer, Véronique. "Misnaming Social Conflict: 'Identity,' Land and Family Histories in a Quilombola Community in the Brazilian Amazon." *Journal of Latin American Studies* 46, no. 3 (2014): 527–55.

Brannstrom, Christian. "Producing Possession: Labour, Law and Land on a Brazilian Agricultural Frontier, 1920–1945." *Political Geography* 20 (2001): 859–83.

Brooke, James. "Brazil Seeks to Return Ancestral Lands to Descendants of Runaway Slaves." *New York Times*, August 15, 1993.

Brown, Ras Michael. *African-Atlantic Cultures and the South Carolina Lowcountry*. Cambridge: Cambridge University Press, 2012.

Brown, Vincent. *The Reaper's Garden: Death and Power in the World of Atlantic Slavery*. Rev. ed. Cambridge, MA: Harvard University Press, 2010.

Bryant, Sherwin K. *Rivers of Gold, Lives of Bondage: Governing through Slavery in Colonial Quito*. Chapel Hill: University of North Carolina Press, 2014.

Butler, Kim D. *Freedoms Given, Freedoms Won: Afro-Brazilians in Post-Abolition São Paulo and Salvador*. New Brunswick, NJ: Rutgers University Press, 1998.

Camp, Stephanie M. H. "'I Could Not Stay There': Enslaved Women, Truancy and the Geography of Everyday Forms of Resistance in the Antebellum Plantation South." *Slavery and Abolition* 23, no. 3 (2002): 1–20.

Campbell, John. "As 'a Kind of Freeman'?: Slaves' Market-Related Activities in the South Carolina Upcountry, 1800–1860." In *The Slaves' Economy: Independent Production by Slaves in the Americas*, edited by Ira Berlin and Philip D. Morgan, 131–69. London: Frank Cass, 1991.

Cancela, Cristina Donza. "Casamento e Relações Familiares na Economia da Borracha (Belém 1870–1920)." Unpublished PhD diss., Universidade de São Paulo, 2006.

Cândido, Antônio. *Os parceiros do Rio Bonito: Estudo sobre o caipira paulista e a transformação dos seus meios de vida*. 11th ed. Rio de Janeiro: Ouro sobre Azul, 2010.

Cardoso, Ciro Flamarion S. "The Peasant Breach in the Slave System: New Developments in Brazil." *Luso-Brazilian Review* 25 (1988): 49–57.

Carney, Judith A. *Black Rice: The African Origins of Rice Cultivation in the Americas*. Cambridge, MA: Harvard University Press, 2002.

———. "Landscapes of Technology Transfer: Rice Cultivation and African Continuities." *Technology and Culture* 37, no. 1 (January 1996): 5–35.

———. "'With Grains in Her Hair': Rice in Colonial Brazil." *Slavery & Abolition* 25, no. 1 (2004): 1–27.

Carney, Judith A., and Richard Nicholas Rosomoff. *In the Shadow of Slavery: Africa's Botanical Legacy in the Atlantic World*. Berkeley: University of California Press, 2009.

Carvalho, Ana Paula Comin de. "O Quilombo da 'Família Silva': Etnicização e politização de um conflito territorial na cidade de Porto Alegre/RS." In *Prêmio ABA/MDA Territórios Quilombolas*, 37–49. Brasília: Ministério do Desenvolvimento Agrário / Núcleo de Estudos Agrários e Desenvolvimento Rural, 2006.

Carvalho, José Jorge de, ed. *O Quilombo do Rio das Rãs: Histórias, Tradições, Lutas*. Salvador: EDUFBA, 1996.

Carvalho, José Murilo de. *The Formation of Souls: Imagery of the Republic in Brazil*. Notre Dame, IN: University of Notre Dame Press, 2012.

———. "Mandonismo, Coronelismo, Clientelismo: Uma Discussão Conceitual." *Dados* 40, no. 2 (1997): 229–50.

Castilho, Celso, and Camillia Cowling. "Funding Freedom, Popularizing Politics: Abolitionism and Local Emancipation Funds in 1880s Brazil." *Luso-Brazilian Review* 47, no. 1 (2010): 89–120.

Castro, Edna. *Escravos e Senhores de Bragança: Documentos Históricos do Século XIX, Região Bragantina, Pará.* Belém: Alves Editora, 2006.

———. *Quilombolas do Pará.* CD-Rom. Belém: UFPA/NAEA, 2005.

Chalhoub, Sidney. *Visões da liberdade: Uma história das últimas décadas da escravidão na corte.* São Paulo: Companhia das Letras, 1990.

Chambouleyron, Rafael. "Cacao, Bark-Clove and Agriculture in the Portuguese Amazon Region in the Seventeenth and Early Eighteenth Century." *Luso-Brazilian Review* 51, no. 1 (2014): 1–35.

———. "Escravos do Atlântico Equatorial: Tráfico Negreiro para o Estado do Maranhão e Pará (Século XVII e Início do Século XVIII)." *Revista Brasileira de História* 26 (2006): 79–114.

———. *Povoamento, Ocupação e Agricultura na Amazônia Colonial (1640–1706).* Belém: Açaí / UFPA, 2010.

Chasteen, John Charles. *National Rhythms, African Roots: The Deep History of Latin American Popular Dance.* Albuquerque: University of New Mexico Press, 2004.

Cleary, David. "'Lost Altogether to the Civilised World': Race and the Cabanagem in Northern Brazil, 1750 to 1850." *Comparative Studies in Society and History* 40 (1998): 109–35.

Clement, Charles R., Michelly De Cristo-Araújo, Geo Coppens D'Eeckenbrugge, Alessandro Alves Pereira, and Doriane Picanço-Rodrigues. "Origin and Domestication of Native Amazonian Crops." *Diversity* 2, no. 1 (2010): 72–106.

Coimbra, Creso. *O Pará na Revolução de 30: História, Análise Crítica e Reflexão.* Belém: Instituto de Cultura, 1981.

Conselho Nacional de Proteção aos Índios, and Ministério da Agricultura. *Diário das Três Viagens (1877–1878–1882) Do Revmo. Padre Nicolino Rodrigues de Sousa Ao Rio Cuminá Afl. Margem Esq. Trombetas Do Rio Amazonas.* Rio de Janeiro: Imprensa Nacional, 1946.

Cordeiro, Paulo. *O Carimbó da Vigia.* Vigia: Edições do Autor, 2010.

Costa, Francisco de Assis. *Ecologismo e Questão Agrária na Amazônia.* Estudos SEPEQ. Belém: Núcleo de Altos Estudos Amazônicos, 1992.

———. "Lugar e significado da gestão pombalina na economia colonial do Grão-Pará." *Nova Economia* 20 (2010): 167–206.

Cottrol, Robert. *The Long, Lingering Shadow: Slavery, Race, and Law in the American Hemisphere.* Athens: University of Georgia Press, 2013.

Courlander, Harold. *A Treasury of Afro-American Folklore: The Oral Literature, Traditions, Recollections, Legends, Tales, Songs, Religious Beliefs, Customs, Sayings, and Humor of Peoples of African Descent in the Americas.* New York: Da Capo Press, 2002.

Coutinho, Amélia, and Sérgio Flaksman. "Barata, Magalhães." In *Dicionário Histórico-Biográfico Brasileiro: 1930–1983,* edited by Israel Beloch and Alzira Alves De Abreu, 1:294–96. Rio de Janeiro: Editora Forense-Universitária / Fundação Getúlio Vargas / Financiadora de Estudos e Projetos, 1984.

Coutinho, Leonardo, Igor Paulin, and Júlia de Medeiros. "A farra da antropologia oportunista." *Veja,* May 5, 2010.

Cowling, Camillia. *Conceiving Freedom: Women of Color, Gender, and the Abolition of Slavery in Havana and Rio de Janeiro*. Chapel Hill: University of North Carolina Press, 2013.

Cruz, Maria Cecília Velasco e. "Puzzling out Slave Origins in Rio de Janeiro Port Unionism: The 1906 Strike and the Sociedade de Resistência dos Trabalhadores em Trapiche e Café." *Hispanic American Historical Review* 86, no. 2 (2006): 205–45.

Cummings, Ronald. "Jamaican Female Masculinities: Nanny of the Maroons and the Genealogy of the Man-Royal." *Journal of West Indian Literature* 21, no. 1/2 (2012/2013): 129–154, 224.

Dávila, Jerry. *Diploma of Whiteness: Race and Social Policy in Brazil, 1917–1945*. Durham, NC: Duke University Press, 2003.

de la Torre, Oscar. "'Are They Really Quilombos'? Black Peasants, Politics, and the Meaning of Quilombo in Present-Day Brazil." *Ofo: Journal of Transatlantic Studies* 3, no. 1–2 (2013): 97–118.

———. "Los ambiguos efectos de la fluidez y la contingencia: la postemancipación en el Brasil fronterizo (Amazonía, 1888–1950)." *Boletín Americanista* 68 (2014): 101–20.

———. "O carimbó e a história social da Grande Vigia, Pará, 1900–1950." *Revista Amazônica* IV, no. 2 (2010): 113–50.

Dean, Warren. "A industrialização durante a República Velha." In *Estrutura de Poder e Economia*, 1:250–83. O Brasil Republicano. São Paulo: Difel, 1977.

———. *Brazil and the Struggle for Rubber: A Study in Environmental History*. Cambridge: Cambridge University Press, 1987.

———. "Latifundia and Land Policy in Nineteenth-Century Brazil." *Hispanic American Historical Review* 51 (1971): 606–25.

———. *With Broadax and Firebrand: The Destruction of the Brazilian Atlantic Forest*. Berkeley: University of California Press, 1997.

De Souza Junior, José Alves. *Tramas Do Cotidiano: Religião, Política, Guerra E Negócios No Grão-Pará Do Setecentos*. Belém: Ed.ufpa, 2012.

Dezemone, Marcus. "A Era Vargas e o Mundo Rural Brasileiro: Memória, Direitos e Cultura Política Camponesa." In *Formas de resistência camponesa: Visibilidade e diversidade de conflitos ao longo da história*, 2:53–72. São Paulo: Editora da UNESP, 2012.

Diacon, Todd A. "Peasants, Prophets, and the Power of a Millenarian Vision in Twentieth-Century Brazil." *Comparative Studies in Society and History* 32 (1990): 488–514.

———. *Stringing Together a Nation: Cândido Mariano Da Silva Rondon and the Construction of a Modern Brazil, 1906–1930*. Durham, NC: Duke University Press, 2004.

Dias, Manuel Nunes. *Fomento e mercantilismo: A Companhia Geral do Grão Pará e Maranhão (1755–1778)*. Vol. 1. 2 vols. Belém: Universidade Federal do Pará, 1970.

Dutra, Maria Vanessa Fonseca. *Direitos Quilombolas: Um Estudo de Impacto da Cooperação Ecumênica*. Rio de Janeiro: KOINONIA Presença Ecumênica e Serviço, 2011.

Eltis, David, and David Richardson. "A New Assessment of the Transatlantic Slave Trade." In *Extending the Frontiers: Essays on the New Transatlantic Slave Trade Database*, 1–60. New Haven, CT: Yale University Press, 2008.

Emmi, Marília Ferreira. *A Oligarquia do Tocantins e o domínio dos castanhais*. Belém: UFPA/NAEA, 1999.

———. *Italianos na Amazônia (1870–1950): Pioneirismo Econômico e Identidade*. Belém: Paká-Tatu, 2008.

Esch, Elizabeth, and David Roediger. "One Symptom of Originality: Race and the Management of Labour in the History of the United States." *Historical Materialism* 17, no. 4 (2009): 3–43.

Esteves, Benedita. "O seringal e a constituição social do seringueiro." In *Processos de constituição e reprodução do campesinato no Brasil*, 1:91–111. São Paulo: UNESP, 2008.

Fabiani, Adelmir. "Os Quilombos Contemporâneos Maranhenses e a Luta Pela Terra." *Estudios Históricos* 2 (2009): 1–19.

Fairbanks, Eve. "The Global Face of Student Protest." *New York Times*, December 11, 2015.

Farfán-Santos, Elizabeth. "'Fraudulent' Identities: The Politics of Defining Quilombo Descendants in Brazil." *Journal of Latin American and Caribbean Anthropology* 20, no. 1 (2015): 110–32.

Faria, Sheila de Castro. "Identidade e comunidade escrava: um ensaio." *Tempo* 11, no. 22 (2007): 122–46.

Fausto, Boris. *A Revolução de 1930: Historiografia e História*. São Paulo: Brasiliense, 1982.

———. *Brasil, de Colonia a Democracia*. Madrid: Alianza Editorial, 1995.

———. "Dos governos militares a Prudente-Campos Sales." In *Estrutura de Poder e Economia*, 1:13–51. O Brasil Republicano. São Paulo: Difel, 1977.

———. *São Paulo na Primeira República*. Vol. 75. Série Estudos. Rio de Janeiro: IUPERJ, 1989.

Fernandes, Florestan. *The Negro in Brazilian Society*. New York: Columbia University Press, 1969.

Ferrara, Miriam Nicolau. *A Imprensa Negra Paulista (1915–1963)*. São Paulo: FFLCH / USP, 1986.

Ferreira, Eliana R. "Guerra sem fim: Mulheres na trilha do direito à terra e ao destino dos filhos (Pará, 1835–1860)." Unpublished PhD diss., Pontifícia Universidade Católica de São Paulo, 2010.

Ferreira, Fred Igor Santiago. "A questão racial no Movimento dos Trabalhadores Rurais Sem Terra: Breve discussão, perspectivas e desafios." In *Anais do V Simpósio Internacional Lutas Sociais na América Latina*, 58–74. Londrina, Brazil, 2013.

Ferreira, Rebeca Campos. "O Artigo 68 do ADCT/CF-88: Identidade e Reconhecimento, Ação Afirmativa, ou Direito Étnico?" *Revista Habitus* 8 (2009): 1–18.

Ferrer Castro, Armando, and Mayda Costa Alegre. *Fermina Gómez y la casa olvidada de Olokun*. Havana: Ediciones Cubanas, 2012.

Fick, Carolyn E. *The Making of Haiti: The Saint-Domingue Revolution from Below*. Knoxville: University of Tennessee Press, 1990.

Figueiredo, Vítor Fonseca, and Camila Gonçalves Silva. "Família, latifúndio e poder: As bases do coronelismo no Norte de Minas Gerais durante a Primeira República (1889–1930)." *Dialogos* 16, no. 3 (2012): 1051–84.

Florentino, Manolo Garcia. "The Slave Trade, Colonial Markets, and Slave Families in Rio de Janeiro, Brazil, ca. 1790–ca. 1830." In *Extending the Frontiers: Essays on the New Transatlantic Slave Trade Database*, edited by David Eltis and David Richardson, 275–312. New Haven: Yale University Press, 2008.

Florentino, Manolo, and José Roberto Góes. *A paz das senzalas: famílias escravas e tráfico atlântico, Rio de Janeiro, c. 1790–c. 1850*. Civilização Brasileira, 1997.

Fontes, Edilza Joana Oliveira. "Cultura e política dos anos trinta no Brasil e as memórias do interventor do Pará, Magalhães Barata (1930–1935)." *Revista Estudos Políticos*, no. 7 (2013): 131–51.

Forret, Jeff. *Race Relations at the Margins: Slaves and Poor Whites in the Antebellum Southern Countryside*. Rev. ed. Baton Rouge: Louisiana State University Press, 2010.

Fraga Filho, Walter. *Encruzilhadas da liberdade: histórias de escravos e libertos na Bahia, 1870–1910*. Campinas, SP, Brasil: Editora UNICAMP, 2006.

Frank, Zephyr. "Elite Families and Oligarchic Politics on the Brazilian Frontier." *Latin American Research Review* 36 (2001): 49–74.

Freitas, Décio. *La Revolución de las Clases Infames*. Buenos Aires: El Ateneo, 2008.

French, Jan Hoffman. *Legalizing Identities: Becoming Black or Indian in Brazil's Northeast*. Chapel Hill: University of North Carolina Press, 2009.

Freyre, Gilberto. *The Masters and the Slaves: (Casa-Grande & Senzala) A Study in the Development of Brazilian Civilization*. 2nd ed. New York: Alfred A. Knopf, 1956.

Frikel, Protásio. *Os Kaxúyana: Notas etno-históricas*. Belém: Museu Paraense Emílio Goeldi, 1970.

Frisch, Michael. *A Shared Authority: Essays on the Craft and Meaning of Oral and Public History*. Albany: State University of New York Press, 1990.

Fromont, Cécile. "Dancing for the King of Congo from Early Modern Central Africa to Slavery-Era Brazil." *Colonial Latin American Review* 22, no. 2 (2013): 184–208.

Fry, Peter, and Carlos Vogt. *Cafundó, a Africa no Brasil: Linguagem e Sociedade*. São Paulo: Editora da Unicamp, 1996.

Fuente, Alejandro de la. "Slave Law and Claims-Making in Cuba: The Tannenbaum Debate Revisited." *Law and History Review* 22, no. 2 (June 2004): 339–369.

Funes, Eurípedes. "Mocambos do Trombetas: Memória e etnicidade (séculos XIX e XX)." In *Os Senhores dos Rios: Amazônia, Margens e Histórias*, edited by Mary Del Priore, 227–57. Rio de Janeiro: Elsevier, 2003.

———. "'Nasci nas Matas, Nunca Tive Senhor': História e Memória dos Mocambos do Baixo Amazonas." PhD diss., FFLCH / USP, 1995.

Funes Monzote, Reinaldo. *From Rainforest to Cane Field in Cuba: An Environmental History since 1492*. Chapel Hill: University of North Carolina Press, 2009.

Furtado, Celso. *The Economic Growth of Brazil: A Survey from Colonial to Modern Times*. Berkeley: University of California Press, 1963.

———. *Formação Econômica do Brasil*. 14th ed. São Paulo: Companhia Editora Nacional, 1976.

Futemma, Célia. "The Use of and Access to Forest Resources: The Caboclos of the Lower Amazon and Their Socio-Cultural Attributes." In *Amazon Peasant Societies in a Changing Environment*, edited by Cristina Adams, Rui Murrieta, Walter Neves, and Mark Harris, 215–37. Dordrecht: Springer Netherlands, 2009. doi:10.1007/978-1-4020-9283-1_10.

Garcia, Graciela Bonassa. "Senhores de terra e intrusos: Os despejos judiciais na campanha rio-grandense oitocentista (Alegrete, 1830–1880)." In *Formas de resistência camponesa: Visibilidade e diversidade de conflitos ao longo da história*, 1:127–51. São Paulo: Editora da UNESP, 2012.

Garcia-Navarro, Lulu. "For Descendants of Brazil's Slaves, a Quest for Land." *Morning Edition.* NPR, February 3, 2014. http://www.npr.org/sections/parallels/2014/02/03 /270874330/descendants-of-brazils-former-slaves-seek-justice.

Garfield, Seth. *In Search of the Amazon: Brazil, the United States, and the Nature of a Region.* Durham, NC: Duke University Press, 2014.

Giesen, James C. "'The Truth about the Boll Weevil': The Nature of Planter Power in the Mississippi Delta." *Environmental History* 14, no. 4 (2009): 683–704.

Girardi, Luisa Gonçalves. "'Gente do Kaxuru': Mistura e transformação entre um povo indígena karib-guianense." PhD diss., Universidade Federal de Minas Gerais, 2011.

Gomes, Flavio, Jonas Marçal De Queiroz, and Mauro Cézar Coelho. *Relatos de Fronteiras: Fontes para a História da Amazônia.* Belém: UFPA / NAEA, 1999.

Gomes, Flavio dos Santos. *A Hidra e os Pântanos: Mocambos, Quilombos e Comunidades de Fugitivos no Brasil (séculos XVII–XIX).* São Paulo: UNESP, 2005.

———. "Etnicidade e Fronteiras Cruzadas nas Guianas (sécs. XVIII–XX)." *Estudios Afroamericanos Virtual,* 2004, 30–58.

———. *Nas Terras do Cabo Norte: Fronteiras, Colonização, e Escravidão na Guiana Brasileira, Sécs. XVIII / XIX.* Belém: UFPA, 1999.

———. "Roceiros, Mocambeiros e as Fronteiras da Emancipação no Maranhão." In *Quase-Cidadão: Histórias e Antropologias da Pós-Emancipação no Brasil,* edited by Olívia Maria da Cunha, 147–71. Rio de Janeiro: Editora FGV, 2007.

———. "'A Safe Haven:' Runaway Slaves, Mocambos, and Borders in Colonial Amazonia, Brazil." *Hispanic American Historical Review* 82 (2002): 469–98.

———. "Slavery, Black Peasants, and Post-Emancipation Society in Brazil (Nineteenth Century Rio de Janeiro)." *Social Identities* 10 (2004): 735–56.

Gomes, Flávio dos Santos, and Olívia Maria da Cunha. "Introdução: Que cidadão? Retóricas da igualdade, cotidiano da diferença." In *Quase-cidadão: Histórias e antropologias da pós-emancipação no Brasil,* 7–19. Rio de Janeiro: Editora FGV, 2007.

Gomes, Flavio dos Santos, and Maria Helena Machado. "Migraciones, desplazamientos y campesinos negros en São Paulo y Río de Janeiro (Brasil) en el siglo XIX." *Boletín Americanista* 68 (2014): 15–35.

Goulart, José Alipio. *O Regatão (Mascate Fluvial da Amazônia).* Rio de Janeiro: Conquista, 1968.

Graham, Sandra Lauderdale. *Caetana Says No: Women's Stories from a Brazilian Slave Society.* New York: Cambridge University Press, 2002.

Grandin, Greg. *Fordlandia: The Rise and Fall of Henry Ford's Forgotten Jungle City.* New York: Metropolitan Books, 2009.

Greider, Thomas, and Lorraine Garkovich. "Landscapes: The Social Construction of Nature and the Environment." *Rural Sociology* 59, no. 1 (1994): 1–24.

Guimarães, Alberto Passos. *Quatro séculos de latifúndio.* 3rd ed. Rio de Janeiro: Paz e Terra, 1969.

Guimarães, Elione Silva. *Múltiplos Viveres de Afrodescendentes na Escravidão e no Pós-Emancipação: Família, Trabalho, Terra e Conflito (Juiz de Fora—MG, 1828-1928).* São Paulo: Annablume / Funalfa Edições, 2006.

———. "Rompendo o silêncio: Conflitos consuetudinários e litigiosos em terras pró-indivisas (Juiz de Fora, Minas Gerais—Século XIX)." In *Formas de resistência camponesa: Visibilidade e diversidade de conflitos ao longo da história*, 1:103–26. São Paulo: Editora da UNESP, 2012.

Gusmão, Neusa Maria Mendes de. "Herança Quilombola: Negros, Terras e Direitos." In *Brasil: Um país de Negros?*, edited by Jefferson Bacelar and Carlos Caroso, 143–61. Rio de Janeiro: PALLAS / CEAO, 1999.

Gutiérrez, Bernardo. "Negros con título de propiedad: Descendientes de esclavos negros huidos viven en una de las áreas más intactas y africanas de la Amazonia." *Público*, May 9, 2008.

Hahn, Steven. *A Nation under Our Feet: Black Political Struggles in the Rural South, from Slavery to the Great Migration*. Cambridge, MA: Belknap Press of Harvard University Press, 2003.

Hambly, Wilfrid D. *The Ovimbundu of Angola*. Vol. XXI. Anthropological Series 2. Chicago: Field Museum of Natural History, 1934.

Harris, Mark. *Life on the Amazon: The Anthropology of a Peasant Village*. London: British Academy, 2001.

———. *Rebellion on the Amazon: The Cabanagem, Race, and Popular Culture in the North of Brazil, 1798–1840*. Cambridge: Cambridge University Press, 2011.

———. "'What It Means to Be Caboclo': Some Critical Notes on the Construction of Amazonian Caboclo Society as an Anthropological Object." *Critique of Anthropology* 18, no. 1 (1998): 83–95.

Hawthorne, Walter. *From Africa to Brazil: Culture, Identity, and an Atlantic Slave Trade, 1600–1830*. New York: Cambridge University Press, 2010.

———. "From 'Black Rice' to 'Brown': Rethinking the History of Risiculture in the Seventeenth- and Eighteenth-Century Atlantic." *American Historical Review* 115, no. 1 (February 1, 2010): 151–63.

Heath, Barbara J., and Lori A. Lee. "Memory, Race, and Place." *History Compass* 8, no. 12 (2010): 1352–68.

Hecht, Susanna B. "Factories, Forests, Fields and Family: Gender and Neoliberalism in Extractive Reserves." *Journal of Agrarian Change* 7 (2007): 316–47.

———. *The Scramble for the Amazon and the "Lost Paradise" of Euclides da Cunha*. Chicago: University of Chicago Press, 2013.

Hemming, John. *Amazon Frontier: The Defeat of the Brazilian Indians*. London: Macmillan London, 1987.

Heredia, Beatriz M. Alasia de. "O campesinato e a plantation: A história e os mecanismos de um processo de expropriação." In *Processos de constituição e reprodução do campesinato no Brasil*, 1:39–67. São Paulo: UNESP, 2008.

Hermann, Jacqueline. "Religião e política no alvorecer da República: os movimentos de Juazeiro, Canudos e Contestado." In *O tempo do liberalismo excludente: Da Proclamação da República à Revolução de 1930*, 1:121–60. O Brasil Republicano. Rio de Janeiro: Civilização Brasileira, 2003.

Heuman, Gad. *Out of the House of Bondage: Runaways, Resistance, and Marronage in Africa and the New World*. London: Frank Cass, 1986.

Holston, James. "Restricting Access to Landed Property." In *Insurgent Citizenship: Disjunctions of Democracy and Modernity in Brazil*, 112–45. Princeton, NJ: Princeton University Press, 2008.

Homma, Alfredo Kingo Oyama, and Antônio José Elias Amorim de Menezes. "Avaliação de uma Indústria Beneficiadora de Castanha-do-Pará, na Microrregião de Cametá, PA." Comunicado Técnico. Belém: Embrapa Amazônia Oriental, 2008.

Hutchinson, Harry William. *Village and Plantation Life in Northeastern Brazil*. Seattle: University of Washington Press, 1957.

Ianni, Octavio. *A luta pela terra: História social da terra e da luta pela terra numa área da Amazônia*. Petrópolis: Vozes, 1978.

Ildone, José. *Noções de História da Vigia*. Belém: Edições CEJUP, 1991.

Instituto Nacional de Colonização e Reforma Agrária. "Perguntas e Respostas da Regularização Quilombola." *INCRA*. www.incra.gov.br/sites/default/files/incra -processosabertos-quilombolas-v2.pdf. Accessed May 15, 2016.

———. "Relação de processos de regularização abertos no INCRA." *INCRA*. Accessed May 15, 2016. www.incra.gov.br/sites/default/files/incra-processosabertos-quilombolas -v2.pdf.

IPHAN. *Dossiê: Carimbó*. Inventário Nacional de Referências Culturais. Belém: IPHAN, 2013.

Izard, Miquel. *Orejanos, Cimarrones, y Arrochelados*. Barcelona: Sendai, 1988.

Johnson, Walter. "On Agency." *Journal of Social History* 37 (2003): 113–24.

Kelly-Normand, Arlene Marie. "Africanos na Amazônia: Cem Anos Antes da Abolição." *Cadernos Centro de Filosofia e Ciências Humanas* 18 (1988): 1–21.

Kenny, Mary Lorena. "The Contours of Quilombola Identity in the Sertao." *Luso-Brazilian Review* 50, no. 1 (2013): 140–64.

Klein, Herbert S., and Francisco Vidal. Luna. *Escravismo no Brasil*. São Paulo: EDUSP, 2010.

———. *Slavery in Brazil*. Cambridge: Cambridge University Press, 2009.

Kraay, Hendrik. "Slavery, Citizenship and Military Service in Brazil's Mobilization for the Paraguayan War." *Slavery & Abolition* 18, no. 3 (December 1, 1997): 228–56.

Lacerda, Franciane Gama. *Migrantes cearenses no Pará: faces da sobrevivência, 1889–1916*. Belém: Editora Açaí, 2010.

Lamounier, Bolívar. "Formação de um pensamento político autoritário na Primeira República: Uma interpretação." In *Sociedade e Instituições*, 2:343–74. História geral da civilização brasileira. Rio de Janeiro: Bertrand Brasil, 1997.

Landers, Jane. "Cimarrón and Citizen: African Ethnicity, Corporate Identity, and the Evolution of Free Black Towns in the Spanish Circum-Caribbean." In *Slaves, Subjects, and Subversives: Blacks in Colonial Latin America*, edited by Jane Landers and Barry Robinson, 111–45. Albuquerque: University of New Mexico Press, 2006.

———. "Gracia Real de Santa Teresa de Mose: A Free Black Town in Spanish Colonial Florida." *American Historical Review* 95, no. 1 (1990): 9–30.

La Rosa Corzo, Gabino. *Runaway Slave Settlements in Cuba: Resistance and Repression*. Chapel Hill: University of North Carolina Press, 2003.

Larrea, Cristina, and José Luis Ruíz-Peinado. "Memoria y Territorio Quilombola en Brasil." *Quaderns de l'Institut Català d'Antropologia* 20 (2004): 191–215.

Laviña, Javier. "Comunidades Afroamericanas: Identidad de Resistencia." *Boletín Americanista* 48 (1998): 139–51.

———. *Cuba: plantación y adoctrinamiento*. Santa Cruz de Tenerife: Ediciones IDEA, 2007.

Leite, Ilka Boaventura. "O Projeto Político Quilombola: Desafios, Conquistas e Impasses Atuais." *Estudos Feministas* 16 (2008): 965–77.

———. "The Trans-Historical, Juridical-Formal, and the Post-Utopian Quilombo," 1–17. Manchester: Centre for Latin American and Caribbean Studies (University of Manchester), 2007.

Lentz, Mark W. "Black Belizeans and Fugitive Mayas: Interracial Encounters on the Edge of Empire, 1750–1803." *The Americas* 70, no. 4 (2014): 645–75.

Levine, Robert M. *Father of the Poor? Vargas and His Era*. New York: Cambridge University Press, 1998.

———. *Vale of Tears: Revisiting the Canudos Massacre in Northeastern Brazil, 1893–1897*. Berkeley: University of California Press, 1992.

———. *The Vargas Regime: The Critical Years, 1934–1938*. New York: Columbia University Press, 1970.

Lewis, Earl. "Connecting Memory, Self, and the Power of Place in African American Urban History." *Journal of Urban History*. 21, no. 3 (1995): 347–71.

Lima, Luciano Demetrius Barbosa. "Os Motins Políticos de Um Ilustrado Liberal: História, Memória e Narrativa na Amazônia em Fins do Século XIX." Unpublished MA thesis, Universidade Federal do Pará, 2010.

Linhares, Luiz Fernando do Rosário. "Kilombos of Brazil: Identity and Land Entitlement." *Journal of Black Studies* 34 (2004): 817–37.

Little, Paul E. *Amazonia: Territorial Struggles on Perennial Frontiers*. Baltimore, MD: Johns Hopkins University Press, 2001.

Lobato, Eládio. *Caminho de Canoa Pequena*. Belém, 2007.

Lockley, Timothy James. *Maroon Communities in South Carolina: A Documentary Record*. Columbia: University of South Carolina Press, 2009.

Lopes, Raimundo Helio. "A Delegacia Militar do Norte e o Governo Provisório: disputas políticas e a nomeação dos interventores nortistas." In *Anais do XXVI Simpósio Nacional de História*, 1–17. São Paulo, 2011.

Love, Joseph LeRoy. *The Revolt of the Whip*. Stanford, CA: Stanford University Press, 2012.

Machado, Maria Helena. "De rebeldes a fura-greves: As duas faces da experiência da liberdade dos quilombolas do Jabaquara na Santos da pós-emancipaçao." In *Quase-Cidadão: Histórias e Antropologias da Pós-Emancipação no Brasil*, 241–82. Rio de Janeiro: Editora UFRJ, 1994.

———. *O plano e o pânico: os movimentos sociais na década da abolição*. Rio de Janeiro: Editora UFRJ, 1994.

MacLachlan, Colin. "African Slave Trade and Economic Development in Amazonia." In *Slavery and Race Relations in Latin America*, edited by Robert B. Toplin, 112–45. Westport, CT: Greenwood Press, 1974.

Maestri, Mário, and Adelmir Fabiani. "O mato, a roça e a enxada: a horticultura quilombola no Brasil escravista (séculos XVI–XIX)." In *Formas de resistência camponesa: visibilidade e diversidade de conflitos ao longo da história*, 63–83. São Paulo: Universidade Estadual Paulista, 2008.

Mahony, Mary Ann. "Creativity under Constraint: Enslaved Afro-Brazilian Families in Brazil's Cacao Area, 1870–1890." *Journal of Social History* 41 (2008): 633–66.

———. "The World Cacao Made: Society, Politics, and History in Southern Bahia, Brazil, 1822–1919." PhD diss., 1996.

Malcher, Jane Aparecida Marques, and Maria Ataide. *Territórios Quilombolas*. Vol. 3. Cadernos ITERPA. Belém: ITERPA, 2009.

Malighetti, Roberto. "Identitarian Politics in the Quilombo Frechal: Live Histories in a Brazilian Community of Slave Descendants." *Outlines—Critical Practice Studies*, 2010, 97–112.

Mallon, Florencia E. *Peasant and Nation: The Making of Postcolonial Mexico and Peru*. Berkeley: University of California Press, 1995.

Mann, Charles C., and Susanna B. Hecht. "Brazil's Maroon People." *National Geographic*, April 2012.

Marcondes, Renato Leite. "Fontes censitárias brasileiras e posse de cativos na década de 1870." *Revista de Indias* 71, no. 251 (2011): 231–58.

Marks, Stuart A. *Southern Hunting in Black and White: Nature, History, and Rituals in a Carolina Community*. Princeton, NJ: Princeton University Press, 1991.

Marques, Fernando Luiz Tavares. "Modelo da Agroindústria Canavieira Colonial no Estuário Amazônico: Estudo Arqueológico de Engenhos dos Séculos XVIII e XIX." PhD diss., Pontifícia Universidade Católica do Rio Grande do Sul, 2004.

Marques, Fernando Luiz Tavares, and Scott Douglas Anderson. "Engenhos movidos a maré no estuário do Amazonas: Vestígios encontrados no município de Igarapé-Miri, Pará." *Boletim do Museu Paraense Emílio Goeldi—Série Antropologia* 8 (1992): 295–301.

Marques, Leonardo. *Por aí e por muito longe: Dívidas, migrações e os libertos de 1888*. Rio de Janeiro: Apicuri, 2009.

Marull, Yana. "Tres millones de descendientes de esclavos viven aún en Brasil en remotas aldeas sin ningún servicio." *El Periódico*, July 14, 2009.

Mata, Iacy Maia. "'Libertos de Treze de Maio' e Ex-Senhores na Bahia: Conflitos no Pós-Abolição." *Afro-Ásia* 35 (2007): 163–98.

Mattos, Hebe Maria. *Das Cores do Silêncio: Os Significados da Liberdade no Sudeste Escravista, Brasil Século XIX*. Rio de Janeiro: Arquivo Nacional, 1995.

———. "Identidade camponesa, racialização e cidadania no Brasil monárquico: O caso da 'Guerra dos Marimbondos' em Pernambuco a partir da leitura de Guillermo Palacios." *Almanack Braziliense: Revista Electônica*, no. 3 (2006): 40–46.

———. "Políticas de reparação e identidade coletiva no meio rural: Antônio Nascimento Fernandes e o quilombo São José." *Revista Estudos Históricos* 1, no. 37 (2006): 167–89.

Maxwell, Kenneth. *Pombal: Paradox of the Enlightenment*. New York: Cambridge University Press, 1995.

McAdam, Doug. "Conceptual Origins, Current Problems, Future Directions." In *Comparative Perspectives on Social Movements: Political Opportunities, Mobilizing Structures, and Cultural Framings*, 23–40. Cambridge: Cambridge University Press, 1996.

———, ed. *Political Process and the Development of Black Insurgency, 1930–1970*. 2nd ed. Chicago: University Of Chicago Press, 1999.

McNeill, John R. "The Ecological Atlantic." In *The Oxford Handbook of the Atlantic World: 1450–1850*, 289–304. New York: Oxford University Press, 2011.

———. "Envisioning an Ecological Atlantic, 1500–1850." *Nova Acta Leopoldina* 114, no. 390 (2013): 21–33.

———. *Mosquito Empires: Ecology and War in the Greater Caribbean, 1620–1914*. New York: Cambridge University Press, 2010.

Meggers, Betty. *Amazonia: Man and Culture in a Counterfeit Paradise: Revised Edition*. Washington, DC: Smithsonian Institution Press, 1996.

Meira, Clóvis. *Barata, no centenário de nascimento*. Belém: Imprensa, 1989.

Meira, Sílvio. "Aquisição da propriedade pelo usucapião." *Revista de informação legislativa* 22, no. 88 (1985): 195–228.

Merchant, Carolyn. *American Environmental History: An Introduction*. New York: Columbia University Press, 2007.

Metcalf, Alida C. *Family and Frontier in Colonial Brazil: Santana de Parnaíba, 1580–1822*. Austin: University of Texas Press, 2005.

Miki, Yuko. "Fleeing into Slavery: The Insurgent Geographies of Brazilian Quilombolas (Maroons), 1880–1881." *The Americas* 68, no. 4 (2012): 495–528.

Milner, Neal. "Ownership Rights and the Rites of Ownership." *Law & Social Inquiry* 18 (1993): 227–53.

Minchinton, W. E. "Early Tide Mills: Some Problems." *Technology and Culture* 20, no. 4 (October 1, 1979): 777–86.

Mintz, Sidney W. *Caribbean Transformations*. New York: Columbia University Press, 1989.

Mintz, Sidney W., and Richard Price. *O Nascimento da Cultura Afroamericana: Uma Perspectiva Antropológica*. Rio de Janeiro: PALLAS / CEAB, 2003.

Miranda, Victorino Coutinho Chermont de. *A Família Chermont: Memória Histórica e Genealógica*. Rio de Janeiro, 1982.

Moreira, Edma Silva. "Memória Social e Luta pela Preservação dos Recursos Naturais: o Caso de São João, uma Comunidade Varzeira da Amazônia." In *No Mar, nos Rios e na Fronteira: Faces do Campesinato no Pará*, edited by Jean Hébette, 111–30. Belém: UFPA, 2002.

Moreira, Edma Silva, and Jean Hébette. "Metamorfoses de um campesinato nos Baixo Amazonas e Baixo Xingu paraenses." In *Diversidade do campesinato: expressões e categorias*, 1. Construções identitárias e sociabilidades, 187–207. São Paulo: Editora UNESP; Núcleo de Estudos Agrários e Desenvolvimento Rural, 2009.

Morgan, Philip D. *Slave Counterpoint: Black Culture in the Eighteenth-Century Chesapeake and Lowcountry*. Chapel Hill, N.C.: Omohundro Institute of Early American History and Culture, 1998.

———. "Work and Culture: The Task System and the World of Lowcountry Blacks, 1700 to 1880." *William and Mary Quarterly* 39 (1982): 564–99.

Mori, Scott A., and Ghillen T. Prance. "Taxonomy, Ecology, and Economic Botany of the Brazil Nut (Bertholletia Excelsa Humb. & Bonpl.: Lecythidaceae)." *Advances in Economic Botany* 8 (1990): 130–50.

Motta, Márcia Maria Menendes. "Classic Works of Brazil's New Rural History: Feudalism and the Latifundio in the Interpretations of the Left (1940/1964)." *Historia Crítica*, no. 51 (2013): 121–44.

———. *Nas fronteiras do poder: Conflito e direito à terra no Brasil do século XIX*. Rio de Janeiro: Vício de Leitura / Arquivo Público do Estado do Rio de Janeiro, 1998.

———. "Posseiros no Oitocentos e a construção do mito invasor no Brasil (1822–1850)." In *Formas de resistência camponesa: Visibilidade e diversidade de conflitos ao longo da história*, 1:85–101. São Paulo: Editora da UNESP, 2012.

Motta-Maués, Maria Angélica. *"Trabalhadeiras" e "Camarados": Relações de Gênero, Simbolismo e Ritualização numa Comunidade Amazônica*. Belém: UFPA, 1993.

Moura, Margarida Maria. *Os deserdados da terra: A lógica costumeira e judicial nos processos de expulsão e invasão da terra camponesa no sertão de Minas Gerais*. Rio de Janeiro: Bertrand Brasil, 1988.

Naro, Nancy. "Customary Rightholders and Legal Claimants to Land in Rio de Janeiro, Brazil, 1870–1890." *The Americas* 48 (1992): 485–517.

———. *A Slave's Place, a Master's World: Fashioning Dependency in Rural Brazil*. London: Continuum, 2000.

Neeleman, Gary, Rose Neeleman, and Wade Davis. *Tracks in the Amazon: The Day-to-Day Life of the Workers on the Madeira-Mamoré Railroad*. Salt Lake City: University of Utah Press, 2013.

Nobrega, Aglaer A., Marcio H. Garcia, Erica Tatto, Marcos T. Obara, Elenild Costa, Jeremy Sobel, and Wildo N. Araujo. "Oral Transmission of Chagas Disease by Consumption of Açaí Palm Fruit, Brazil." *Emerging Infectious Diseases* 15, no. 4 (April 2009): 653–55.

Nugent, Stephen. *Amazonian Caboclo Society: An Essay on Invisibility and Peasant Economy*. Providence, RI: Berg, 1993.

———. "Whither O Campesinato? Historical Peasantries of Brazilian Amazonia." *Journal of Peasant Studies* 29 (2002): 162–89.

O'Dwyer, Eliane Cantarino. "DaMatta nas paradas entre 'malandros' ou 'heróis': A lenda da cobra-grande, o tempo histórico e questões de identidade." In *O Brasil Não é Para Principiantes: Carnavais, Malandros e Heróis, 20 Anos Depois*. Rio de Janeiro: Editora Fundação Getúlio Vargas, 2002.

———. "Os Quilombos do Trombetas e do Erepecurú-Cuminá." In *Quilombos: Identidade Étnica e Territorialidade*, edited by Eliane Cantarino O'Dwyer, 255–80. Rio de Janeiro: ABA / FGV Editora, 2002.

———. *Quilombos: Identidade Étnica e Territorialidade*. Rio de Janeiro: Editora Fundação Getúlio Vargas, 2002.

Oliveira Filho, João Pacheco de. "O Caboclo e o Brabo: Notas sobre duas modalidades de força-de-trabalho na expansão da fronteira Amazônica no século XIX." *Civilização Brasileira* (1979): 101–40.

Ortiz, Fernando. *Cuban Counterpoint: Tobacco and Sugar*. Durham, NC: Duke University Press, 1995.

Pace, Richard. "The Amazon Caboclo: What's in a Name?" *Luso-Brazilian Review* 34, no. 2 (1997): 81–89.

Palacios, Guillermo. *Cultivadores libres, Estado y crisis de la esclavitud en Brasil en la época de la Revolución industrial*. Mexico, D.F.: El Colegio de México / Fideicomiso Historia de las Américas / Fondo de Cultura Económica, 1998.

Palheta, Aércio. *Vigia Ainda Ontem*. Belém: IOE, 1995.

Paoliello, Renata Medeiros. "As comunidades tradicionais no Vale do Ribeira: da 'reprodução camponesa' às re-significações dos patrimônios territoriais." *Agrária*, no. 3 (2005): 58–82.

———. "'Condição camponesa' e novas identidades entre remanescentes de quilombos no Vale do Ribeira de Iguape." In *Diversidade do campesinato: expressões e categorias*, 1. Construções identitárias e sociabilidades, 229–50. São Paulo: Editora UNESP / Núcleo de Estudos Agrários e Desenvolvimento Rural, 2009.

Parés, Luis Nicolau. "O Processo de Crioulização no Recôncavo Baiano (1750–1800)." *Afro-Ásia*, no. 33 (2005): 87–132.

Parker, Eugene P. *The Amazon Caboclo: Historical and Contemporary Perspectives*. Vol. 32. Studies in Third World Societies. Williamsburg, VA: Department of Anthropology, College of William and Mary, 1985.

———. "The Amazon Caboclo: An Introduction and Overview." In *The Amazon Caboclo: Historical and Contemporary Perspectives*, edited by Eugene P. Parker, 32:xvii–li. Williamsburg, VA: Department of Anthropology, College of William and Mary, 1985.

Peck, Gunther. "The Nature of Labor: Fault Lines and Common Ground in Environmental and Labor History." *Environmental History* 11, no. 2 (2006): 212–38.

Penningroth, Dylan C. *The Claims of Kinfolk: African American Property and Community in the Nineteenth-Century South*. Chapel Hill: University of North Carolina Press, 2003.

———. "The Claims of Slaves and Ex-Slaves to Family and Property: A Transatlantic Comparison." *American Historical Review* 12 (2007): 1039–69.

Pereira de Carvalho, Fábio. "Autonomia e hierarquia em duas comunidades de escravos: os casos George (Alabama, Estados Unidos, 1847) e Lino (Vassouras, Brasil, 1871)." *Mundo agrario* 13, no. 26 (June 2013): 1–23.

Pereira, Karen Signori, Flávio Luis Schmidt, Rodrigo L. Barbosa, Ana M. A. Guaraldo, Regina M. B. Franco, Viviane L. Dias, and Luiz A. C. Passos. "Transmission of Chagas Disease (American Trypanosomiasis) by Food." *Advances in Food and Nutrition Research* 59 (2010): 63–85.

Peres, Carlos A., and Claudia Baider. "Seed Dispersal, Spatial Distribution and Population Structure of Brazilnut Trees (Bertholletia Excelsa) in Southeastern Amazonia." *Journal of Tropical Ecology* 13 (1997): 595–616.

Petit, Pere. *A Esperança Equilibrista: A Trajetória do PT no Pará*. Belém: Boitempo Editora, 2003.

———. *Chão de Promessas: Elites Políticas e Transformações Econômicas no Estado do Pará pós-1964*. Belém: Paka-Tatu Ltda, 2003.

Piñero, Eugenio. *The Town of San Felipe and Colonial Cacao Economies*. Philadelphia: American Philosophical Society, 1994.

Pinto, Benedita Celeste de Moraes. *Nas Veredas da Sobrevivência: Memória, Gênero e Símbolos de Poder Feminino Em Povoados Amazônicos*. Belém: Paká-Tatu, 2004.

Planas, Roque. "Brazil's 'Quilombo' Movement May Be the World's Largest Slavery Reparations Program." *Huffington Post*, July 10, 2014. www.huffingtonpost.com/2014/07/10/brazil-quilombos_n_5572236.html.

Portelli, Alessandro. *The Death of Luigi Trastulli and Other Stories: Form and Meaning in Oral History*. Albany: State University of New York Press, 1991.

————. "History-Telling and Time: An Example from Kentucky." *Oral History Review* 20, no. 1/2 (1992): 51–66.

Prado Jr., Caio. *The Colonial Background of Modern Brazil*. Berkeley: University of California Press, 1967.

Price, Richard. *First-Time: The Historical Vision of an Afro-American People*. Baltimore, MD: Johns Hopkins University Press, 1983.

————. *Maroon Societies: Rebel Slave Communities in the Americas*. Baltimore, MD: Johns Hopkins University Press, 1979.

Price, Richard, Valdélio Santos Silva, Jean François Véran, and Sheila Brasileiro. "Dossiê Remanescentes de Quilombos." *Afro-Ásia*, no. 23 (1999). www.redalyc.org/toc.oa?id =770&numero=6185.

Priore, Mary del, and Renato Pinto Venâncio. *Uma história da vida rural no Brasil*. Rio de Janeiro: Ediouro, 2006.

Querino, Manoel. "O colono preto como fator da civilização brasileira." *Afro-Ásia*, no. 13 (1980): 143–58.

Raffles, Hugh. *In Amazonia: A Natural History*. Princeton, NJ: Princeton University Press, 2002.

Ramos, Jair de Souza. "Ciência e racismo: Uma leitura crítica de Raça e assimilação em Oliveira Vianna." *História, Ciências, Saúde-Manguinhos* 10, no. 2 (2003): 573–601.

Reis, Arthur Cézar Ferreira. *História de Óbidos*. Brasília: Civilização Brasileira, 1945.

Reis, João José. *A morte é uma festa: Ritos fúnebres e e revolta popular no Brasil do século XIX*. São Paulo: Companhia das Letras, 1990.

————. *Slave Rebellion in Brazil: The Muslim Uprising of 1835 in Bahia*. Baltimore, MD: Johns Hopkins University Press, 1993.

Reis, João José, and Flávio dos Santos Gomes, eds. *Liberdade por um Fio: História dos Quilombos no Brasil*. São Paulo: Companhia das Letras, 1996.

Reis, Nathacha Regazzini Bianchi. "Motins Políticos, de Domingos Antonio Raiol: Memória e historiografia." *Revista Intellectus* 1 (2005): 1–10.

Resende, Maria Efigênia Lage de. "O processo político na Primeira República e o liberalismo oligárquico." In *O tempo do liberalismo excludente: Da Proclamação da República à Revolução de 1930*, edited by Jorge Luiz Ferreira and Lucília de Almeida Neves Delgado, 1:89–121. O Brasil Republicano. Rio de Janeiro: Civilização Brasileira, 2003.

Ribeiro, Vanderlei Vazelesk. "Cartas ao Presidente Vargas: Outra Forma de Luta pela Terra." In *Formas de resistência camponesa: Visibilidade e diversidade de conflitos ao longo da história*, 2:53–72. São Paulo: Editora da UNESP, 2012.

Ricci, Magda. "De la Independencia a la Revolución Cabana: La Amazonía y el Nacimiento de Brasil (1808–1840)." In *La Amazonía Brasileña en Perspectiva Histórica*, edited by José Manuel Santos, 59–91. Salamanca: Ediciones Universidad de Salamanca, 2006.

Rios, Ana Maria Lugão. "'My Mother Was a Slave, Not Me!': Black Peasantry and Regional Politics in Southeast Brazil, 1870–1940." PhD diss., University of Minnesota, 2002.

Rios, Ana Maria Lugão, and Hebe Maria Mattos. *Memórias do cativeiro: Família, trabalho e cidadania no pós-abolição*. Rio de Janeiro: Civilização Brasileira, 2005.

————. "Para além das senzalas: Campesinato, política e trabalho rural no Rio de Janeiro pós-Abolição." In *Quase-cidadão: Histórias e antropologias da pós-emancipação no Brasil*, 55–78. Rio de Janeiro: Editora FGV, 2007.

Ritchie, Donald A. *Doing Oral History: A Practical Guide*. 2nd ed. Oxford: Oxford University Press, 2003.

Rivera-Barnes, Beatriz. "Ethnological Counterpoint Fernando Ortiz and Jean Price-Mars, or Santeria and Vodou." *SAGE Open* 4, no. 2 (2014): 1–11.

Rocque, Carlos. *Magalhães Barata: O Homem, a lenda, o político*. Vol. 1. Belém: SECULT, 1999.

Rodrigues, João Lucas. "Serra dos Pretos: Trajetórias de famílias egressas do cativeiro no pós-abolição (Sul de Minas, 1888–1950)." *Afro-Ásia*, no. 50 (2014): 171–97.

Rodrigues, José Damião. "'Para o Socego e Tranquilidade Publica das Ilhas:' Fundamentos, Ambição e Limites das Reformas Pombalinas nos Açores." *Tempo* 11 (2006): 144–70.

Rodrigues, Nina. *Os Africanos no Brasil*. Edited by Homero Pires. São Paulo: Editora Nacional, 1977.

Rogers, Thomas D. *The Deepest Wounds: A Labor and Environmental History of Sugar in Northeast Brazil*. Chapel Hill: University of North Carolina Press, 2010.

Roller, Heather Flynn. "Colonial Collecting Expeditions and the Pursuit of Opportunities in the Amazonian Sertão, C. 1750–1800." *The Americas* 66 (2010): 435–67.

————. "Colonial Routes: Spatial Mobility and Community Formation in the Portuguese Amazon." PhD diss., Stanford University, 2010.

Rubert, Rosane Aparecida, and Paulo Sérgio da Silva. "O acamponesamento como sinônimo de aquilombamento: O amálgama entre resistência racial e resistência camponesa em comunidades negras rurais do Rio Grande do Sul." In *Diversidade do campesinato: expressões e categorias*, Vol. 1. Construções identitárias e sociabilidades, edited by Emilia Pietrafesa de Godoi et al, 251–74. São Paulo: Editora UNESP; Núcleo de Estudos Agrários e Desenvolvimento Rural, 2009.

Ruiz-Peinado, José Luis. "Amazonía Negra." In *La Amazonía Brasileña En Perspectiva Histórica*, edited by José Manuel Santos Pérez and Pere Petit, 23–59. Salamanca: Ediciones Universidad de Salamanca, 2006.

————. *Cimarronaje en Brasil: Mocambos del Trombetas*. Valencia, Spain: El Cep i la Nansa, 2003.

————. "El Empadronamiento de los Dioses: Indios y Negros en la Amazonia Colonial." In *Mitos Religiosos Afroamericanos: Cultura y Desarrollo*, 67–83. Barcelona: Centre d'Estudis i Recerques Socials i Metropolitanes, 2014.

————. "Maravilla, Ataque y Defensa de un Mocambo en la Amazonía." In *Relaciones Sociales e Identidades en América*, edited by Gabriela Dalla Corte, 107–21. Barcelona: Edicions de la UB, 2002.

————. "Misioneros en el río Trombetas: La subida del Padre Carmelo de Mazzarino." *Boletín Americanista* 54 (2004): 177–98.

————. "Tiempos Afroindígenas en la Amazonia: Primera Mitad del Siglo XIX." *Revista de Indias* LXX (2010): 583–607.

Rupert, Linda M. "Marronage, Manumission and Maritime Trade in the Early Modern Caribbean." *Slavery & Abolition* 30, no. 3 (2009): 361–82.

Saes, Décio. *Classe média e política na Primeira República Brasileira (1889–1930)*. Petrópolis: Vozes, 1975.

Salles, Vicente. "Memória sobre a rede de dormir que fazem as mulheres índias e negras no Grão-Pará, conforme anotações de cronistas antigos e modernos." In *O Negro na Formação da Sociedade Paraense: Textos Reunidos*, 101–12. Belém: Paká-Tatu, 2004.

———. *O Negro no Pará sob o regime da escravidão*. Rio de Janeiro: Fundação Getúlio Vargas-Universidade Federal do Pará, 1971.

———. *Os Mocambeiros e outros ensaios*. Belém: IAP, 2013.

Salles, Vicente, and Marena Isdebski Salles. "Carimbó: Trabalho e lazer do Caboclo." *Revista Brasileira de Folklore* 9 (1969): 257–82.

Salomão, Rafael de Paiva. "Densidade, Estrutura e Distribuição Espacial de Castanheira-do-Brasil (Bertholletia excelsa H. & B.) em Dois Platôs de Floresta Ombrófila Densa na Amazônia Setentrional Brasileira." *Boletim do Museu Paraense Emílio Goeldi. Ciências Naturais* 4 (2009): 11–25.

Santos, Roberto. *História econômica da Amazônia: 1800–1920*. São Paulo: T. A. Queiroz, 1980.

Sauma, Julia. "Ser Coletivo, Escolher Individual: Território, medo e família nos Rios Erepecurú e Cuminã," 1–21. Caxambu: ANPOCS, 2009.

Schmitt, Alessandra, Maria Cecília Manzoli Turatti, and Maria Celina Pereira De Carvalho. "A Atualização do Conceito de Quilombo: Identidade e Território nas Definições Teóricas." *Ambiente & Sociedade*, 2002, 1–6.

Schmutzer, Kurt. "Der Liebe Zur Naturgeschichte Halber Johann Natterers Reisen in Brasilien 1817–1835." PhD diss., University of Vienna, 2007.

Schwartz, Stuart. *Slaves, Peasants, and Rebels: Reconsidering Brazilian Slavery*. Urbana: University of Illinois Press, 1992.

———. *Sugar Plantations in the Formation of Bahian Society: Bahia, 1550–1835*. Cambridge: Cambridge University Press, 1985.

Scott, Rebecca. "Reclamando la Mula de Gregoria Quesada: El Significado de la Libertad en los Valles del Arimao y del Caunao, Cienfuegos, Cuba." *Illes i Imperis* 2 (1999): 89–108.

Scott, Rebecca, and Michael Zeuske. "Demandas de Propiedad y Ciudadanía: Los Exesclavos y sus Descendientes en la Región Central de Cuba." *Illes i Imperis* 5 (Fall 2001): 109–34.

SEPPIR, Secretaria Especial de Políticas de Promoção da Igualdade Racial. *Programa Brasil Quilombola*. Brasília: SEPPIR, 2005.

Shepard Jr., Glenn H., and Henri Ramirez. "'Made in Brazil': Human Dispersal of the Brazil Nut (Bertholletia Excelsa, Lecythidaceae) in Ancient Amazonia." *Economic Botany* 65, no. 1 (2011): 44–65.

Silva, Daniel B. Domingues da. "The Atlantic Slave Trade to Maranhão, 1680–1846: Volume, Routes and Organisation." *Slavery & Abolition* 29, no. 4 (2008): 477–501.

Silva, Francisco Carlos Teixeira da, and Maria Yedda Linhares. *Terra Prometida: Uma história da questão agrária no Brasil*. Rio de Janeiro: Campus, 1999.

Silver, Timothy. *A New Face on the Countryside: Indians, Colonists, and Slaves in South Atlantic Forests, 1500–1800*. Cambridge: Cambridge University Press, 1990.

Simonsen, Roberto. *História econômica do Brasil: 1500–1820*. Vol. 34. Edições do Senado Federal. Brasília: Senado Federal, Conselho Editorial, 2005.

Sistema de Monitoramento do Programa Brasil Quilombola, Fundação Cultural Palmares. "Quadro geral de Comunidades Remanescentes de Quilombos (CRQs)." *Fundação Cultural Palmares*, December 31, 2015. www.palmares.gov.br/wp-content/uploads/2016 /01/tABELA_CRQs_COMPLETA-Atualizada-31-12-15.pdf.

Skidmore, Thomas E. *Brasil: De Getúlio a Castelo (1930–1964)*. Rio de Janeiro: Editôra Saga, 1969.

Slater, Candace. *Entangled Edens: Visions of the Amazon*. Berkeley: University of California Press, 2002.

Slenes, Robert W. "'Malungu, ngoma vem!': África coberta e descoberta do Brasil." *Revista USP*, no. 12 (1992): 48–67.

———. *Na senzala, uma flor: Esperanças e recordações na formação da família escrava— Brasil Sudeste, Século XIX*. Rio de Janeiro: Editora Nova Fronteira, 1999.

Smith, Nigel J. H. *The Amazon River Forest: A Natural History of Plants, Animals, and People*. New York: Oxford University Press, 1999.

———. *The Enchanted Amazon Rain Forest: Stories from a Vanishing World*. Gainesville: University Press of Florida, 1996.

Sodré, Emmanuel. *Lauro Sodré na História da República*. Rio de Janeiro: Edição do Autor, 1970.

Soluri, John. *Banana Cultures: Agriculture, Consumption, and Environmental Change in Honduras and the United States*. Austin: University of Texas Press, 2005.

Souza, Bruno. "Nos Limites da Justiça Paraense: Estrutura Agrária e Conflitos na Região do Tocantins—1860 a 1880." MA thesis, Universidade Federal do Pará, 2013.

Stanfield, Michael Edward. *Red Rubber, Bleeding Trees: Violence, Slavery, and Empire in Northwest Amazonia 1850–1933*. Albuquerque: University of New Mexico Press, 1998.

Stédile, João Pedro, and Bernardo Mançano Fernandes. *Brava Gente: El MST y la lucha por la tierra en el Brasil*. Barcelona: Virus, 2002.

Stein, Stanley J. *Vassouras: A Brazilian Coffee County, 1850–1900: The Roles of Planter and Slave in a Plantation Society*. Princeton, NJ: Princeton University Press, 1985.

Stevenson, Brenda E. "The Question of the Slave Female Community and Culture in the American South: Methodological and Ideological Approaches." *Journal of African American History* 92, no. 1 (2007): 74–95.

Steward, Julian H., ed. *Handbook of South American Indians*. Vol. 3. Bureau of American Ethnology Bulletin 148. Washington, DC: Smithsonian Institution, 1948.

Stewart, Mart A. "Slavery and the Origins of African American Environmentalism." In *"To Love the Wind and the Rain": African Americans and Environmental History*, 9–20. Pittsburgh, PA: University of Pittsburgh Press, 2006.

———. *"What Nature Suffers to Groe": Life, Labor, and Landscape on the Georgia Coast, 1680–1920*. Wormsloe Foundation Publications: No. 19. Athens: University of Georgia Press, 1996.

Sweet, James H. "Defying Social Death: The Multiple Configurations of African Slave Family in the Atlantic World." *William and Mary Quarterly* 70, no. 2 (2013): 251–72.

———. *Recreating Africa: Culture, Kinship, and Religion in the African-Portuguese World, 1441–1770*. Chapel Hill: University of North Carolina Press, 2003.

Tarrow, Sidney. *Power in Movement: Social Movements and Contentious Politics.* 2nd ed. Cambridge: Cambridge University Press, 1998.

Tavares, João Walter. *Inventário Cultural, Social, Político e Econômico de Oriximiná.* Oriximiná: Prefeitura Municipal de Oriximiná, 2006.

Teixeira, Heloísa Maria. "Família escrava, sua estabilidade e reprodução em Mariana, Minas Gerais, 1850–1888." *Afro-Ásia,* no. 28 (2002): 179–220.

Teixeira, Lygia Conceição Leitão. *Marambiré: O Negro no Folclore Paraense.* Belém: Secretaria de Cultura / Fundação Cultural do Pará Tancredo Neves, 1989.

Telles, Edward. *Race in Another America: The Significance of Skin Color in Brazil.* Princeton, NJ: Princeton University Press, 2004.

Thompson, Alvin O. *Flight to Freedom: African Runaways and Maroons in the Americas.* Jamaica: University of the West Indies Press, 2006.

Tilley, Christopher. "Introduction: Identity, Place, Landscape and Heritage." *Journal of Material Culture* 11, no. 1–2 (2006): 7–32.

Tocantins, Sylvia Helena. *No tronco da Sapopema: Vivências interioranas.* Belém: Imprensa Oficial, 1998.

Toplin, Robert B. *The Abolition of Slavery in Brazil.* New York: Atheneum, 1975.

Toral, André Amaral de. "A participação dos negros escravos na Guerra do Paraguai." *Estudos Avançados* 9 (1995): 287–96.

Treccani, Girolamo Domenico. *Terras de Quilombo: Caminhos e Entraves do Processo de Titulação.* Belém: Programa Raízes, 2006.

———. *Violência e Grilagem: Instrumentos de Aquisição da Propriedade da Terra no Pará.* Belém: UFPA/ITERPA, 2001.

Tucker, Richard P. *Insatiable Appetite: The United States and the Ecological Degradation of the Tropical World.* Berkeley: University of California Press, 2000.

Turner, Mary. *From Chattel Slaves to Wage Slaves: The Dynamics of Labour Bargaining in the Americas.* Indianapolis: Indiana University Press, 1995.

Vansina, Jan. *Oral Tradition as History.* Madison: University of Wisconsin Press, 1985.

———. *Paths in the Rainforests: Toward a History of Political Tradition in Equatorial Africa.* Madison: University of Wisconsin Press, 1990.

Vasconcelos, Marcus Arthur Marçal de, Ruy Rangel Galeão, Ana Vânia Carvalho, and Valéria Nascimento. "Práticas de Colheita e Manuseio do Açaí." *Embrapa Amazônia Oriental—Documentos,* no. 251 (2006).

Véran, Jean-François. "Quilombos and Land Rights in Contemporary Brazil." *Cultural Survival* 25, no. 4 (2010): 20–25.

———. "Rio das Rãs: Memória de uma 'comunidade remanescente de quilombo.'" *Afro-Ásia,* no. 23 (1999): 295–323.

Wagley, Charles. *Amazon Town: A Study of Man in the Tropics.* New York: Alfred A. Knopf, 1968.

Walsh, John C., and Steven High. "Rethinking the Concept of Community." *Histoire Sociale / Social History* 32, no. 64 (1999): 255–73.

Watts, David. *The West Indies: Patterns of Development, Culture and Environmental Change since 1492.* Cambridge: Cambridge University Press, 1990.

Weimer, Rodrigo de Azevedo. *Os nomes da liberdade: ex-escravos na serra gaúcha no pós-abolição.* Sao Leopoldo: Oikos Editora, 2008.

Weinstein, Barbara. *The Amazon Rubber Boom: 1850–1920*. Stanford, CA: Stanford University Press, 1983.

———. "Persistence of Caboclo Culture in the Amazon: The Impact of the Rubber Trade, 1850–1920." In *The Amazon Caboclo: Historical and Contemporary Perspectives*, edited by Eugene P. Parker, 32:89–113. Williamsburg, VA: Department of Anthropology, College of William and Mary, 1985.

Welch, Clifford Andrew, Edgard Malagodi, Josefa S. B. Cavalcanti, and Maria de Nazareth B. Wanderley, eds. *Camponeses brasileiros*. Vol. 1. Leituras e interpretações clássicas. História Social do Campesinato no Brasil. São Paulo: Editora UNESP, 2009.

Werner, Alice. *Myths and Legends of the Bantu*. London: George W. Harrap and Co., 1933.

Witkoski, Antonio Carlos. *Terras, florestas e águas de trabalho: os camponeses amazônia e as formas de uso de seus recursos naturais*. São Paulo: Annablume, 2010.

Woodard, James P. "Coronelismo in Theory and Practice: Evidence, Analysis, and Argument from São Paulo." *Luso-Brazilian Review* 42, no. 1 (2005): 99–117.

———. *A Place in Politics: São Paulo, Brazil, From Seigneurial Republicanism to Regionalist Revolt*. Durham, NC: Duke University Press, 2009.

Yabeta, Daniela, and Flavio Gomes. "Memória, cidadania e direitos de comunidades remanescentes (em torno de um documento da história dos quilombolas da Marambaia)." *Afro-Ásia*, no. 47 (2013): 79–117.

Young, Allen M. *The Chocolate Tree: A Natural History of Cacao*. Washington, DC: Smithsonian Institution Press, 1994.

Yow, Valerie Raleigh. *Recording Oral History: A Guide for the Humanities and Social Sciences*. 2nd ed. New York: Altamira Press, 2005.

Zuidema, Pieter A., and René G. A. Boot. "Demography of the Brazil Nut Tree (Bertholletia Excelsa) in the Bolivian Amazon: Impact of Seed Extraction on Recruitment and Population Dynamics." *Journal of Tropical Ecology* 18, no. 1 (2002): 1–31

Index

abolition of slavery, 6, 26, 44–45, 46, 75, 101–2

Abuí community. *See* Lake Abuí

açaí (*Euterpe oleracea*), 21, 46, 58, 98, 107

Acará River, 2, 20, 23, 29, 55, 63; Cabanagem in the, 25–26

Africa: slaves from, 9, 10–14 passim, 26, 28, 31, 36, 56, 59; subsistence strategies from, 54, 56–58, 60; traditions from, 7–8, 14, 37–38, 49, 65–66, 72, 76, 93, 106

agouti (*Dasyprocta*), 55, 98, 104

agricultural knowledge. *See* Africa: subsistence strategies from

Alenquer: Brazil nut trade in, 61, 78, 84–86, 88, 116–21 passim, 130; city council of, 118; history of, 34–35, 117–18, 120–21; Italian community of, 121–22, 130; land fraud in, 82; livestock ranching in, 77; local elites of, 88, 119–21, 133; Magalhães Barata's visit to, 130; maroon-descendants in, 144; marronage in, 36, 41, 44; peasant protests in, 118–19, 140; police violence in, 85–86, 130; state visit to, 1, 116, 118–19, 123–26 passim

Alexandre (maroon-descendant leader), 121

American Confederate exiles, 59

Andrade, José Julio de, 129, 131

Andrea, Francisco Soares, 24–25

Angelim (Eduardo Francisco Nogueira), 24

Angola: oral traditions from, 37–38; slaves from, 6, 13–15, 28, 100, 160n18; subsistence strategies from, 57, 106, 139

annatto, 19, 29

ARQMO (Quilombo-Descendant Association of the Municipality of Oriximiná), xii, 33, 147

Article 68, 3–5, 143–45

bacurizeiro (*Platonia insignis*), 19

Baixo Amazonas. *See* Lower Amazon region

banana, 18, 19, 21, 43, 46, 58, 75

Barbosa Rodrigues, João, 47

Barracão de Pedra. See Rocky Shed

Barros, Sebastião do Rego, 40, 44

Bates, Henry Walter, 17, 19, 25, 51, 54, 55, 70

Batista Campos, João Gonçalves de, 22, 23, 26

beans, 21, 43, 55, 57, 75, 97, 104, 159n12

Belém do Pará: Cabanagem revolt in, 22–24; history of, 12, 14; plantations near, 20, 22, 25–26, 31, 51, 54–57, 70; rivers flowing to, 20, 62; rubber production in, 62; slavery in, 56, 71, 96; slave trade to, 12–14, 28, 136; state and city governments of, 110, 128, 133, 138

Benevides, Francisco de Sá e, 19, 30

Bentes, Dionysio, 124, 126, 128–29

Berthier, Jean-Jacques, 26

Biard, François, 55

Big Snake. *See* Snakes

black peasants: concept of, 9; conflicts of, 85–88, 93–96 passim, 108–11, 116–20 passim, 125–26, 137–38; culture of, 48–50, 76, 104–8 passim; current politics of, 1–4, 33, 115, 133–34; economic activities of, 5, 46–47, 75–76, 82–84, 92 passim, 102–4, 119–20; gender inequality among, 70, 90–91, 103–4; geographic mobility of, 44–48, 75, 77, 86–87, 102–3; identity, 4, 9–10, 48–50, 114–15, 139–41, 144–45; nationalism among, 121–23, 132–33, 140–41; origins of, 5–11 passim, 44, 52, 54–64 passim, 72–73, 100–101, 119–20, 140; patron-client ties among, 86–92 passim, 120–21, 132; political discourses of, 95–96, 108–15 passim, 119–23, 132, 140–41; religion among, 76